ARMY & NAVY ACADEMY

HISTORY OF THE WEST POINT OF THE WEST

ALEXANDER MUI

THE History PRESS

Published by The History Press
Charleston, SC
www.historypress.net

First published 2017

Manufactured in the United States

ISBN 9781467137881

Library of Congress Control Number: 2016961714

Notice: The information in this book is true and complete to the best of our knowledge. It is offered without guarantee on the part of the author or The History Press. The author and The History Press disclaim all liability in connection with the use of this book.

Dedicated in the loving memory of my mother,
as well as Niki Pavelic, Class of 2008.

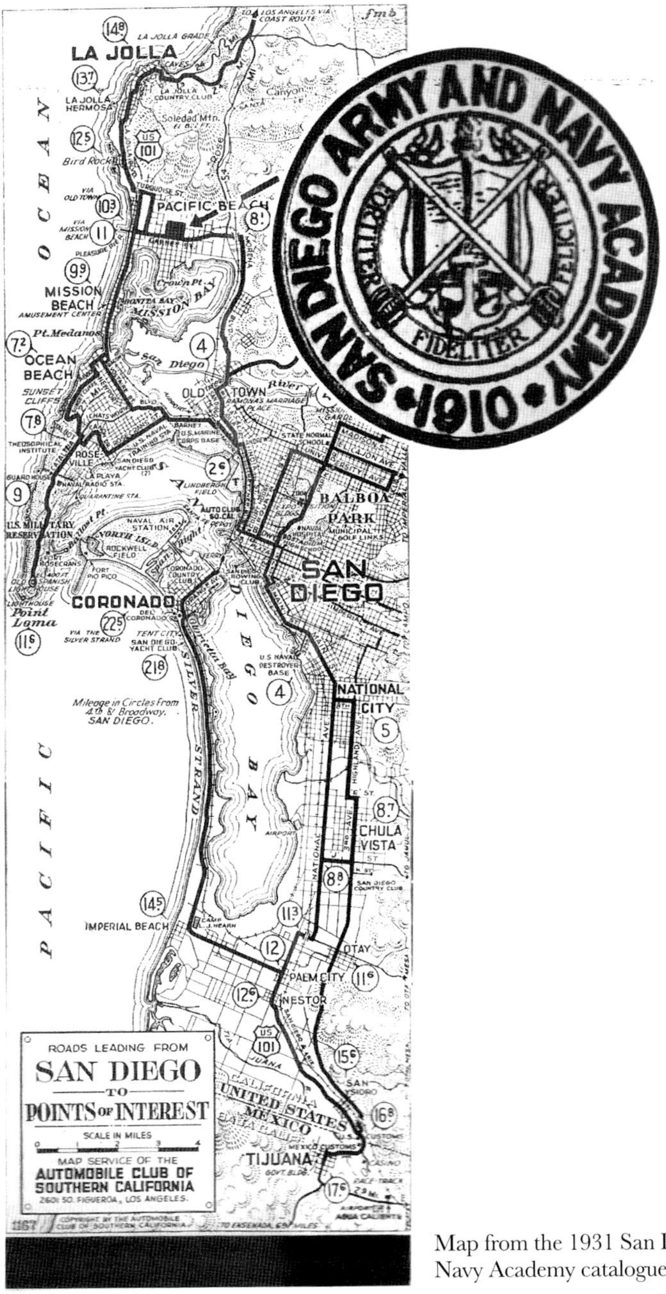

Map from the 1931 San Diego Army and Navy Academy catalogue. *Author's collection.*

Contents

Acknowledgements 7
Author's Note 9

Prologue: Marching Westward
Preface. The End of an Era 13
Introduction. The Military Schools of America 15
1. Sons of the South 23

Part I: By the City's Northern Border
2. Founding of the Academy 35
3. Those Distant Years 49
4. The Lost Academy 59

Part II: The Academy's Golden Age
5. California's West Point 69
6. Conquer and Prevail 78
7. The Black Year 90

Part III: Sheltered by the Sun-Kissed Hills
8. Bordered by the Sea 99
9. The Interregnum 105
10. To Same Heights 109

Contents

Part IV: The Atkinson Era
11. Atkinson's Lament 125
12. Towards the Sunset 134
13. The Golden Anniversary 147

Part V: They Just Fade Away
14. All Glory Is Fleeting 159
15. Strange Sense of Nostalgia 169
16. The Year of Five Presidents 182

Part VI: The Academy's Renaissance
17. Into a New Millennium 193
18. Foundations of the Revival 201
19. Brother Stands by Brother 207

Part VII: Towards the Centennial
20. An Industrial Age 229
21. Centennial Celebrations 235
22. Into a New Century 242

Afterword. Story of a Cadet Lifer, by Bradley D. Larsen 253
Appendix I. Presidents of the Academy 257
Appendix II. Songs of the Academy 259
Appendix III. Flags of the Academy 261
Appendix IV. Seals and Crests of the Academy 263
Additional Source Notes 265
Select Bibliography 283
Index 285
About the Author 288

Acknowledgements

Special acknowledgement goes to Bradley D. Larsen for his support and gracious funding of the publication of this book. This history research could not have been done without the help and support of Susy Smith (ANA Alumni director), Ken Hughey (ANA, 1983; ANA Alumni Association president), Jeffrey Tisor (ANA, 1985), Mark J. Sellinger (ANA), Nicole Ramsey (ANA advancement director), David Falkenau (ANA faculty), Pete Zoschak, Joe Orozco (ANA staff), Stephen Bliss (ANA president), Martin Davis (ANA faculty), Julie Hall (ANA librarian), Emily Scarano (ANA librarian), Louis Bitterlin and Coach John Maffucci (ANA athletic director).

Additional thanks go to Doug Halbert (adjutant of the Brown Military Academy Alumni Association, 1950), Roger Van Rekom (Southern California Military Academy, 1948; ANA, 1951), Gary Moore (Brown Military Academy Glendora, 1963), Johnnie Crean Jr. and Karen Miller. Gratitude is also extended to the alumni I interviewed from the San Diego Army and Navy Academy, Davis Military Academy, Army and Navy Academy, Brown Military Academy (Pacific Beach), Brown Military Academy (Glendora), Harding Military Academy, San Diego Military Academy and Southern California Military Academy.

Grateful acknowledgment is made to the Army and Navy Academy's Alumni Association, Samuel Warfield Peterson Library Archives, William C. Atkinson History Museum and Office of Advancement, as well as the Brown Military Academy Alumni Association and San Diego Military Academy Alumni Association, for use of photographs and images not owned by the author in this publication.

Acknowledgements

A special word of thanks goes to supporters of the 2014 Kickstarter campaign and those who promoted the project, including Bradley D. Larsen (ANA, 1972); Scott H. Biggar (ANA, 1969); Major General Arthur Bartell (ANA president); Dr. Robin Cowen, PhD (ANA faculty chair); Marco Chayet (ANA, 1989); Jason Colwell; Ann Jasper; Mark J. Sellinger (ANA, 1986); Douglas J. Halbert (BMA, 1950); Richard A. McBride (ANA, 1957); Max Isbell (ANA, 1971); David Brain Stiles (ANA, 1972); David Thomas (ANA, 2006); Joe Agosta (ANA, 2008); Sean Hukill; Norman Scherner (ANA, 1989); Walter Caron (BMA, 1968); Tyler Dinelle; Joshua Moyer; Issac J. Hostak; Christopher Russell; Mark Duffer; Ashton Brown (BMA, 1951); Leonid Milman (ANA, 2013); Ethan Segovia (ANA faculty); Dan Loden (ANA faculty); Brock Barry (ANA, 2009); Duran Heath (ANA, 2016); Jacob Hughes (U.S. Army); Ben Haines; Jason D. Knight; Alex Ricks; Richard Lincon Sr. (ANA staff); Bobbi Patterson (ANA staff); Lisa Cafro (ANA faculty); Cori Friedman (ANA faculty); Joe Dreiske (ANA, 1987); Major Jeff Marshburn (ANA, 1990); Brian Maxwell (ANA, 2006); Ryan J. Silverman (ANA, 2007); Eric Wang (ANA, 2007); Kevin Chu (ANA, 2008); Daniel Long (USMC); William L. Frazier; Jonathan Piper; Stephen Farrell; Christina Emily; Kendra and Craig Millner; Charles Bowerman; Kyle Emerick; Andrew "Brutamod" Kepple; Bryan Kincaid; Michael J. Lucio; and Forward Operating Base Equestria.

AUTHOR'S NOTE

All words in between quotations come from a primary source or interview referenced in the source notes. Some information may include additional footnotes describing more details, conflicting stories or author's commentary. Chapters 19 and 20, which cover periods contemporary to the author, are written from an autobiographical perspective.

I will often refer to the Army and Navy Academy and other schools in a personified form using feminine pronouns. The word *Academy* (capitalized) refers to the Army and Navy Academy in all her incarnations. The word *Corps* refers to the Army and Navy Academy's Corps of Cadets of the era being addressed, with the same intention applied to *Alumni*. The word *campus* is used when referring to the Academy's different campuses (e.g., "Pasadena campus" or "Pacific Beach campus"). The word *Board* refers to the Board of Directors of the Academy before 1941 and the Board of Trustees afterwards. Thomas A. Davis is referred to as "Captain Davis" before 1921 and "Colonel Davis" afterwards. His younger brother, John L. Davis Jr., is referred to as "Lieutenant Davis" before 1936 and "Major Davis" afterward. William C. Atkinson is referred to as "Captain Atkinson" before 1941 and "Colonel Atkinson" or "Colonel Atty" afterwards. The name *TAC Officer* refers to Tactical Officer, Tactics Department Staff or Teacher-Advisor-Counselor, depending on military school and era. Throughout the Academy's history, the terms *Headmaster* and *Dean* are used interchangeably referring to the same position as the highest-ranking head in the Academics Department, as are *Commandant* and *Superintendent* for the head of the school (and occasionally as head of the Residential Department) and *Military Instructor* and *Senior Army Instructor* for the head of the Military Education Department. The "High School" is also referred to as the "Upper School," and the "Middle School" is referred to as the "Junior School" and "Lower School."

Abbreviations used include: San Diego Army and Navy Academy (SDANA), Army and Navy Academy (ANA), El Paso Military Institute (EPMI), Brown Military Academy (BMA), San Diego Military Academy (SDMA), Southern California Military Academy (SCMA), Junior Reserve Officers' Training Corps (JROTC) and Association of Military Colleges and Schools of the United States (AMCSUS).

The Army and Navy Academy's history is often divided into three eras: the Golden Years or Davis Era, the Atkinson Era and the Modern Era. I have further divided these periods for this book. For my chronology, the Davis Era has been divided into the Platinum Age (1907–17) and Golden Age (1917–37); the Interregnum (1938–43) covers the period in between the Davis and Atkinson Eras; the Atkinson Era has been divided into the Silver Age (1943–72) and Bronze Age (1972–91); and the Modern Era has been divided into the Dark Age (1991–98), the Renaissance (1998–2008) and the Industrial Age (2009–present).

The Prologue tells the dual story of the post–Civil War birth and westward expansion of the American military school and the Army and Navy Academy's founder, Colonel Thomas A. Davis, along with the founding of the El Paso Military Institute. Part I covers the Academy's Platinum Age, founding of the San Diego Army and Navy Academy in Pacific Beach and the lost story of the sister school in Pasadena. Part II covers the Academy's Golden Age, the rise of military schools in California and the height of the San Diego Army and Navy Academy' success and following downfall. Part III covers the Academy's move to Carlsbad as the Davis Military Academy and the years trying to reclaim former glory. Part IV covers the Academy's Silver Age and the return to success under the guidance of Colonel William C. Atkinson, as well as the takeover of the Pacific Beach campus by Brown Military Academy. Part V covers the Academy's Bronze and Dark Ages, the post–Vietnam War decline of military schools and the stories of the San Diego Military Academy, Brown Military Academy in Glendora and Southern California Military Academy. Part VI covers the Academy's Renaissance and adjustments in a post-9/11 world, as well as the domination of the JROTC program. Part VII covers the Academy's Centennial Celebration and changes made to survive into a second century.

PROLOGUE

MARCHING WESTWARD

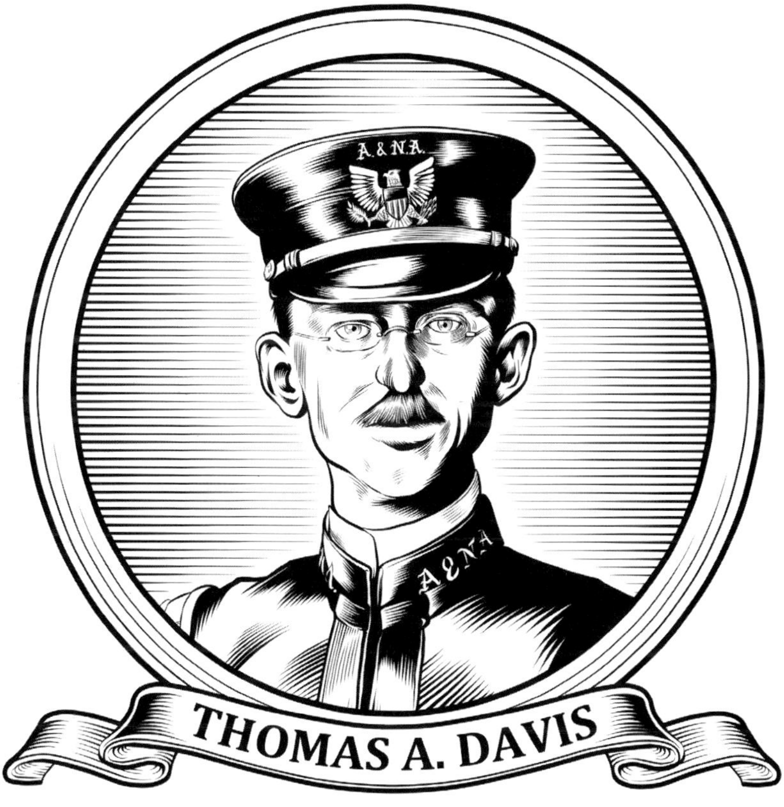

Thomas A. Davis. Artwork by Alexander Mui.

PREFACE
THE END OF AN ERA

I ALWAYS TELL PEOPLE that along with the house I grew up in, the Army and Navy Academy is the only place I call home. What a rich story she has to tell. As a new cadet, my curiosity in wanting to learn the history of the Academy caused me to write this book. What began as a few notes I jotted in my Plebe Book my first year later became a seventy-page research paper submitted for my tenth-grade English class. After nearly a decade of research, this book you are holding now is the fruit of my labor.

The Army and Navy Academy is the last college preparatory military school in California and sole survivor on the West Coast. The Academy was founded back in the day when thousands of military schools dotted this nation, yet the Army and Navy Academy distinguished herself from the rest of them. The founder, Colonel Thomas A. Davis, created a system of education that was revolutionary for its time. Despite disadvantages compared to the many larger, established and financially well-off military schools of that era, the Academy stood above when it came to attracting prospective cadets to whom the Davis style of education appealed. This success aided the Academy during the period when, one by one, military schools across the nation closed down until there was only a handful left.

In fact, in 2010, when I was completing a rough draft of this book in time for the Academy's 100th Centennial Celebration, I learned that yet another military school had been forced to close. The New York Military Academy, a much older school just down the road from West Point, sadly had to close its doors due to low funding, despite having an alumni base that includes

President Donald Trump. This left the Army and Navy Academy just one of twenty-seven military high schools left in the nation. With numerous other military schools having closed since then, it is apparent that we are nearing the end of an era. It's for these reasons that now, more than ever, the need to preserve the history of the Army and Navy Academy is paramount.

While this book is primarily about the Army and Navy Academy, during the writing process I realized that ANA's existence does not solely reside in a bubble. Her story is part of a much larger story about the American military school experience. Therefore, I wanted to step back and take a look at the larger picture. Alongside my research on ANA, I gathered as much information as I could on as many other military schools.

It's a history that fascinates me. While I initially tried to dig up as much information as I could find on any notable military school, either existing or defunct, this proved to be an impossible task. It has been reported that at one point in time, there existed up to 750 military schools. We may never know how many military schools there once were, as the definition of a military school is not easy to define. Many schools that use the title of "military school" include religious boarding schools, discipline reform schools, JROTC-sponsored public day schools and reform schools.

Looking at many different lists of defunct military schools, I gathered that many were not in existence for more than five years before closing. However, the establishment of military schools does appear to follow historical trends—a large majority of them in the 1800s and early 1900s exploded into existence after major wars, just as an equal number fizzled out as quickly as they appeared.

When I started my research back in 2005, no major book on military schools had been published; therefore, for the first few drafts, I was forced to do my own firsthand research and come to my own conclusions independent of other historians. My research started from primary sources such as faded newspaper articles and firsthand interviews. I can proudly say that in the age of the quick Google search, the acquisition of the sources for my book was almost completely Internet free—the only times the Internet was used was to view existing scans of old newspapers and books I could not otherwise get my hands on.

In writing this book, I feel that the most important information that needed to be recorded was the story of the Academy's early history, as there are no surviving alumni from that era. I felt it was imperative to document this forgotten history before it vanishes with time. Therefore, you may notice that the first half of this book covering the San Diego Army and Navy Academy is more in-depth and detailed than the latter half. I feel that the more recent stories of the Academy's history would be better told by the alumni who experienced them.

Introduction

The Military Schools of America

Military schools defined a major part of American culture. Nostalgic imagery of a bugler sounding reveille at sunrise putting into motion the scramble of cadets out of barracks to morning formations, parades of gray uniforms marching to a drum cadence in the afternoon sun and the haunting call of taps sounding off the night. The graduates of military schools include some of the most influential political and military leaders of the modern world, as well as countless leaders of the business, law and entertainment world.

Once seen as the model school to mold young boys into well-mannered and educated gentlemen, society changed and so did the way the nation educated young men. By the 1960s, with the antiwar movement on the rise and following government distrust in the post-Watergate era, military schools quickly diminished in the slice of the pre-college market.

Perhaps part of the decline results from the bad reputation military schools have had in culture. In media, institutions of education are portrayed in different manners: inner-city public schools are portrayed as drug dens rife with gang violence, while preparatory boarding schools are seen as establishments where wealthy, elitist families send their spoiled children. Military schools have perhaps received the most negative portrayals. In fiction, they are seen as reform schools to send juvenile delinquents and misbehaving children. While this is, to an extent, true, it's just one slice of a larger picture. Films like *TAPS*, *Dress Grey* and *Lords of Discipline* have birthed the modern impression of such military schools being responsible for breeding militarized, one-dimensional characters with no other redeeming traits. Real military schools are much more diverse than their one-note fictional counterparts.[1]

The disciplinary reform school that uses a military structure is perhaps the most common interpretation of the military school, although today those are few and far between, if any remain at all. These schools never had the status or accreditation of the college preparatory military academies, and none had a seat at the Association of Military Colleges and Schools of the United States, the sole governing body of accredited military schools.

This conflation is due to the fact that the name "military school" encompasses a broad spectrum. Many educational institutions have branded themselves as military schools, including federal and state-sponsored military institutions, privately owned West Point–style preparatory academies, religious boarding schools, discipline reform schools and even JROTC-sponsored public day schools. A vast majority of military schools established in America are the religious boarding schools, which use the military structure to discipline their students. These religious boarding schools were most often founded as extensions of local churches and were more concerned with teaching faith-based principles and blind obedience than actual real-world education. These schools would often use cadets as free labor at a church-owned farm or mill. These workhouse environments made use of manual labor to instill Christian virtues into students. Corporal punishment was common among these religious boarding schools and still is; when such schools later added "military school" to their names to attract more prominence, these actions were transposed to the wider cultural image of military schools in general.

Hazing is another aspect of military schools featured in fiction that is actually more common in any normal public school, boarding school and even in elite colleges and Ivy League universities. You are more likely to see gang violence and bullying in a public school than a military school, while even more elite boarding schools and universities hold higher rates of suicides and deaths due to hazing. Even my own alma mater, Johns Hopkins University, known as the "Cutthroat School," is plagued with the fact that it has among the highest student suicide rates among undergraduate universities due to stress. Failure to address students' psychological health is often a common attribute at such schools. In more recent news, Johns Hopkins and other universities have faced public backlash for prominent

1. Counterpoints to such portrayals are humorous films like *Up the Academy* and Disney's *Cadet Kelly*, although the cadets in these films are also portrayed as one-dimensional gag characters.

hazing and harassment cultures that exist on these campuses and which are still often given the blind eye by the administration. It is therefore unfair that military schools have become synonymous with such traits when they appear more often in other types of educational institutions. In comparison, the system of accountability and level of attention dedicated to each cadet helps relieve stress, as each cadet is checked and cared for, while character training and leadership structure help greatly decrease hazing, making military college preparatory schools far safer environments.

Another possible reason for the negative portrayal in the cultural landscape is the myth that military schools are trying to turn young children into fighting soldiers instead of providing proper educations. This is not true. In fact, no pre-college military school of the modern era requires its students to enlist in the military upon graduation. In fact, most students of these institutions make their way to elite colleges and universities. In the last decade, the Army and Navy Academy has seen 95 to 100 percent of her graduates make their way to a college or university.

With that said, it is a wonder why pre-college military schools are now nearly extinct in the modern era. There have been a few theories put forth explaining the decline, including the aforementioned reasons and a shifting culture. In the course of my research, I have discovered some new theories, including the improved convenience in travel and the rise of the JROTC Program and its expansion to a wider range of private and public schools. However, before we discuss the military school decline, it is appropriate to look at the birth of this American institution.

THE FIRST MILITARY SCHOOLS

The first American military schools could have been called proto-military schools. Most took a form that has no relation to the military schools of today. The oldest military high school to be tracked down is the Charlotte Hall Military Academy in Maryland, which was founded in 1774. Located in Charlotte Hall, this school was established by Queen Charlotte as Charlotte Hall School to "provide for the liberal and pious education of the youth of this province, the better to fit them for the discharge of their duties." Originally a school that educated only girls, in the 1850s the school was converted into a military boarding school for boys and was later renamed Charlotte Hall Military Academy. After two centuries of operation, the school closed in 1976 due to financial hardship and was converted into the Charlotte Hall Veterans' Home.

This school is indeed older than the federal service academies; however, the original Charlotte Hall School was not a military school until the 1850s. In fact, no military school established before 1800 survived to the twentieth century. Most of these early military schools originated as private boarding schools for boys and would only later pick up military structures—after the formations of the United States Military Academy at West Point and the United States Naval Academy at Annapolis—as a better way of educating boys. Most of these rebranded schools operated primarily as preparatory schools for young men who were hoping to attend the federal military academies at West Point and Annapolis.

The United States Military Academy at West Point is the obvious template for all successive military colleges and schools. President Thomas Jefferson set into motion plans to establish a national military school at West Point, which was a fort first occupied by the Continental army on January 27, 1778, during the American Revolution. Congress authorized the establishment of the United States Military Academy with the Military Peace Establishment Act of 1802, which was signed by Jefferson in March of that year. The academy officially started operations of July 4, 1802. Jonathan Williams served as the first superintendent. During these early years, the academy welcomed cadets as young as ten, who would attend for a period of six months to six years. On the eve of the War of 1812, Congress authorized the foundation of a more formal education system and allowed for the Corps to increase to 250 cadets.

West Point alumnus Captain Alden Partridge, Class of 1806, remained with the Academy after graduation as the assistant professor of mathematics, eventually making his way up the ranks to the position of acting Superintendent in 1808. Partridge transformed West Point into a model of moral perfection by requiring all cadets to attend church service and ensuring that physical fitness was a primary focus. He would often lead the Corps on long marches through New York and neighboring states. "Old Pewt," as he was called, was never immoderate, often delivering the Sunday church sermons. It was like he never left the military school life, often doing physical training alongside cadets while micromanaging his subordinates within the Academy faculty, with most considering him a martinet.

In 1814, Captain Partridge was officially appointed as Superintendent of West Point. During his tenure, there was a shortage of the traditional blue

West Point uniform, and so he had gray uniforms made in New York City to supply the cadets. In 1816, the War Department decided to make this cheaper gray uniform the official uniform of West Point, as "it better suits the finance of the Cadets than one of the blue." Thus began the tradition of the "Long Gray Line."

In 1817, a former student of Partridge's, Colonel Sylvanus Thayer, now his superior officer, was appointed the new Superintendent of West Point. Partridge refused to relinquish his position to a former student. However, one court-martial later, Partridge resigned his commission in 1818. He left the U.S. Army, serving his entire career at West Point.

As soon as Colonel Sylvanus Thayer was appointed as Superintendent, he established what is known as the West Point curriculum, the same one that currently influences all American military schools to this day. Thayer, known as the "Father of the Military Academy," established the Cadet Honor Code, demerit system and a system requiring cadets to be of outstanding military bearing and appearance, as well as numerous other traditions. As this structure was imitated by later military schools, Thayer could likewise be considered the "Father of Military Academies." The first test of this new educational style came when graduates of Thayer's tenure proved themselves in battle during the Mexican-American War. Such distinguished graduates included future Civil War commanders Robert E. Lee and Ulysses S. Grant. This period has been often romanticized by graduates on both sides of the Civil War as the "end of the Old West Point era."

Meanwhile, Captain Alden Partridge decided to establish his own military academy. In Norwich, Vermont, he established the American Literary, Scientific and Military Academy in 1819, thus launching the first of the private military colleges. Now known as Norwich University, this institution is considered the first non-federal military academy and is recognized as the nation's oldest private military college. Within the first years of operation, around 480 students from twenty-two of the then twenty-four states attended this new school. Partridge started the tradition of military education on a civilian campus and promoted the idea of the "citizen solider," a man trained to act in a military capacity in time of war but capable of fulfilling standard civilian functions in time of peace. This idea led to the formation of National Guard and Reserve units, with regimented training, replacing local militia forces. This would eventually pave the way for the Reserve Officers' Training Corps (ROTC) Program, which would become the main competing alternative to the federal service academies in terms of collegiate military training. The United States

Department of Defense officially recognizes Norwich University as the "Birthplace of ROTC."

As West Point and Norwich University flourished during this era, they also inspired the emergence of state-sponsored military colleges, including schools like the Virginia Military Institute (1839) in Lexington, Virginia, and The Citadel (1842) in Charleston, South Carolina.

Virginia Military Institute (VMI) was a hybrid school combining the best characteristics of the United States Military Academy and École Polytechnique in Paris, creating a new system that became the template adopted by many military schools formed in the South. Later, educational grants, state per state, would also allow for the creation of the first military pre-college schools modeled on the state schools. It is notable that most military schools that have come into existence were based in the South, with the highest percentage in Virginia and South Carolina (near VMI, Virginia Tech and The Citadel). In fact, Virginia has the highest number of the surviving military high schools (six as of 2014).

Virginia Military Institute would later gain renown and respect during the Civil War, as it remains the only military school to have enrolled cadets called into active military engagements. VMI authorized fourteen battle streamers for its flag; however, it decided to only carry one, for the Battle of New Market, in which the VMI Corps of Cadets fought as an independent unit. VMI suffered fifty-five casualties, with ten cadets killed in battle. This is the only time in American history that a school's entire student body fought as a unit in battle. General John C. Breckinridge, the commanding Confederate officer, initially put the cadets in reserve. Upon seeing Union forces break the Confederate line, he stated, "Put the boys in…and may God forgive me for that order." The cadets held the line and eventually pushed forwards to capture the Union artillery emplacement, winning a victory for the South. VMI is the only military school in the United States to have bayonets on its drill rifles due to the action of its cadets during this battle.

School for Gentlemen, Boys and God

During the Civil War, most of the newly formed Southern military schools were burned and destroyed by Union forces, while many Northern military schools had to relinquish their land and property to the Union army. Most of these schools would not resume operations after the war. Examples of such schools are Fleetwood Academy (1839–1860) and Georgia Military Institute (1851–1864), which was burned by General Sherman.

However, a fair number of military high schools from this period have survived to this day. Notable schools founded during this era include the Carson Long Military Academy (1836), Marion Military Academy (1842), St. John's College High School (1851) and Oak Ridge Military Academy (1852).

The next and largest surge of military school formation came after the Civil War, as many retired officers decided to set up military boarding schools to serve as easy income during their retirement, like Augusta Military Academy (1865) in Fort Defiance, Virginia, founded by Confederate veteran Charles Roller. These schools were marketed as upper-class institutions to refine young boys into gentlemen and focused more on personal grooming and discipline than the vision of the reformatory military school. Schools like Augusta would even reject or expel any students who displayed less than impeccable behavior. However, as most of these schools were created as a source of a simple paycheck for retired officers, they would not survive the deaths of their founders. The average military school from this period would cease to exist in only five years.

However, as the large volume of military schools were created during this post–Civil War boom, nearly half of the surviving military schools today were founded during this era. Sixteen notable schools are still in operation, including Georgia Military College (1879), Fishburne Military School (1879), Wentworth Military Academy (1880), New York Military Academy (1889), Missouri Military Academy (1889) and New Mexico Military Institute (1891).

As America was founded as a bastion of religious freedom, so too were the early schools fundamentally concerned with religion. Before the United States was a young republic, such religious schools were the first to be formed in the colonies, including the Sabbath primers of New England, the Catholic schools of Maryland and the Quaker institutions of Pennsylvania. The earliest colleges were all religiously based. Harvard University started in 1636 as a Puritan school to train ministers, just as Princeton was founded by Presbyterians and Yale by Congregationalists, while the University of Pennsylvania was founded by Methodists with a trust charity requested by evangelist George Whitefield.

Religious families of the time saw it as their duty to properly model their sons into moral men who were meant to be Christian warriors. In this same tradition, many of the military schools formed after the bloody Civil War were religious schools founded by churches, friars, monks, nuns and other faith-based organizations. Predominately Roman Catholic and Episcopal, these schools were established originally with no specific affiliation to the military, yet they adopted the structure of the military schools created by the retired Civil War officers as a backbone to manage their students. The military structure was very similar to the religious hierarchy of these faiths and thus lent itself well to such repurposing. For example, St. John's Military School (1887) was founded as the Episcopal Military Institute by the Episcopal Diocese of Kansas, and Randolph-Macon Academy (1892) was chartered by the United Methodist Church, which administered the admission and curriculum (which was primarily religiously based).

Indeed, most of the military schools that survive to this day were founded as religious schools with a military structure, including Christian Brothers Academy (1859), St. John's Northwestern (1884), Howe Military School (1884), Saint Thomas Academy (1885), St. Catherine's Academy (1889), Texas Military Institute (1893) and Benedictine Military School (1902).

One significant event that allowed for the formation of new civilian military colleges and cadet corps was the Land Grant Act of 1862. During the Civil War, this act, known as the Morrill Act, led to the creation of new schools by allowing for the establishment of colleges that specifically taught courses in military tactics. Under this act, each eligible state received thirty thousand acres of federal land. Colleges that were created or benefited from this act were required to have military training as part of their curriculum—some made it mandatory for all male students. These land grant college military tactic programs are considered the forerunners of the current ROTC system, which cites the 1862 Morrill Act as being the birth of the program.

After the war, the Morrill Act was extended to southern states. In 1867, Congress passed a special act granting Tennessee eligibility to participate in the Morrill Act. The Eastern Tennessee University in Knoxville was designated the state's first recipient of the grant funds. During the war, the university was right in the middle of a divided conflict, as Tennessee was a Confederate state but populated by Union sympathizers. As Confederate and Union troops tried to occupy the city, the campus of the university was forced to close its doors to students right before the Siege of Knoxville broke out. During the siege, the campus's buildings were turned into barracks and hospitals. The campus, not far from Fort Sanders, was severely damaged during the Battle of Fort Sanders. With the grants from the Morrill Act, the school could recover and established its own cadet corps. In 1879, it was renamed the University of Tennessee. One alumnus who graduated from the university's cadet corps was a young man from Knoxville named Thomas A. Davis.

Chapter 1
Sons of the South

On a cold February afternoon, Thomas Alderson Davis's body laid in repose at Green Mortuary in San Diego. Present at his funeral service were William C. Atkinson, president of the Army and Navy Academy, and Louis Bitterlin, president of the San Diego Military Academy. It was no coincidence that these two individuals were the heads of the last two military high schools in California. Only separated by fourteen miles, it is also notable that these last two educational institutions in the state were both located in the San Diego area. These two represented the foreboding end of the West Coast military school tradition, one that had been started half a century earlier by the very man lying in the coffin.

This one event, this final meeting of these three men—only mentioned fleetingly in brief local obituaries—marked the end of a chapter in the little-known history of military education in America.

In his prime, Thomas Alderson Davis was hailed as a "significant figure" in the shaping of the history of Southern California, an adapter of the West Point style in a high school structure, a progressive mastermind in new ways of educating America's young men, a founder of military schools, a president of the largest military academy west of the Mississippi and head of a transnational conglomerate of schools and colleges. However, by the time of his passing in 1964, his name had faded from history, and his only remaining legacy was one lone military high school on the beach of Carlsbad.

In time, his influence would be all but forgotten even within his last institution. A painting done by English artist Arnold Mountfort in the 1930s was lost, and no portrait of the late colonel exists anywhere on this last campus. His name is now known only as a piece of trivia for plebes and as the namesake of the Thomas A. Davis Administration Hall, which serves as the center of campus.

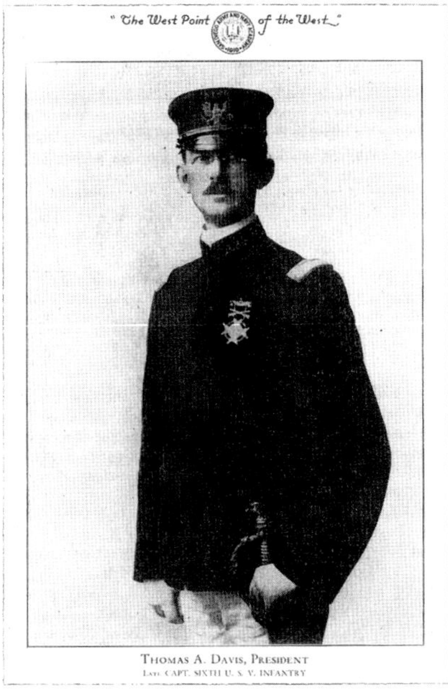

Portrait of Captain Thomas A. Davis, circa 1912. *Author's collection.*

Like any founder of an educational institution, his life's story has been rewritten to serve whatever role he needs to play. When Brown Military Academy wanted to lay claim to the heritage and achievements of the San Diego Army and Navy Academy, Colonel Davis was offered a job at his old campus and pushed to the forefront of every promotional catalogue. When I was enrolled as a cadet at the Army and Navy Academy in 2004, the story told was that the founder of the "West Point of the West" was, in fact, a top West Point graduate. However, he never attended the United States Military Academy. Recently, during the Centennial rebranding of the Academy, the newly proposed "Six Pillars" of the school were credited as a creation of the late colonel.

Much is forgotten about this man the American Historical Society described in 1921 as an influential resident in shaping the growth of the San Diego and Los Angeles area. He was a man who created a system for the education of America's young men that predated the Boy Scouts and ROTC; founded one of the only military schools that was non-religious and welcomed students of all faiths, cultures and nationalities, from Mexico to China, at a time when such progressive inclusiveness was not welcomed; hired women to teach and hold prominent positions of power; and pushed for new ways to educate students, not just with lectures in the classroom but with outdoor lessons and hands-on activities in the fresh air, where classes were divided by mental maturity not age. Here was a man with a lifelong dream, one that perhaps he never felt was fully realized.

Thomas Alderson Davis

Thomas Alderson Davis was born in Lebanon, Virginia, on June 29, 1873, in a South just recovering from the American Civil War. Thomas was the son of John Lynch Davis Sr. and Mary Jennet "Mollie" Alderson Davis. He had two younger brothers: Charles H. Davis, born in August 1875, and John Lynch Davis Jr., born in June 1892.

Not much is known of his childhood. Publications speaking of Thomas mention him being raised in a traditional southern household of Christians and military gentlemen. He attended public school in Lebanon until the age of eight, when his family moved to Knoxville, Tennessee. There Thomas attended various grammar schools before entering high school.

During his high school years, Thomas played as a pitcher on his school's baseball team, was the top student in his class, was a member of the Glee Club and played cello in the orchestra. In 1888, Thomas entered the preparatory department of the University of Tennessee.

In 1889, Thomas enrolled into the University of Tennessee as a full-time student. The university benefited from grants allowed by the Morrill Act, although in return it had to allow for the training of military tactics and formed its own cadet corps. This was something that intrigued Thomas, who joined the military tactics program right away. As a poor native of Knoxville, this would be the closest he could get to a West Point–style education. Since a military structure in an academic environment greatly interested Thomas, he considered a career in education. The following year, Thomas became a member of the National Guard of Tennessee and served as the senior captain of the university's battalion from 1890 to 1892. While in university, he was inducted as a member of Phi Gamma Delta fraternity. In 1893, Thomas graduated with his bachelor degree in liberal arts.

The Spanish-American War

Thomas A. Davis joined his father in the wholesale shoe business in Knoxville after graduation. This would become his normal routine until the Spanish-American War broke out in 1898. Seeking a change of scenery, Davis joined the 6th Volunteer Infantry. On June 29, 1898, he was commissioned as a first lieutenant and quickly rose through the ranks. Davis was promoted to regimental adjutant on December 1 of that same year. He was stationed in

Puerto Rico during the Puerto Rico Campaign. On January 7, 1899, Davis was promoted to captain of his company.

This war would influence the way Davis would reinterpret his own upbringing. He strongly felt that it was the manner in which his family raised him, as a gentleman and a leader, that prepared him for this war. Caught up in the spirit of the time, he felt that all young Americans should be better prepared in coming to the defense of their nation. Davis believed that all young men should have the kind of upbringing he had, yet he knew that not every boy had such an opportunity. This ignited a new vision for Davis: a military-structured school that would help raise young boys to be the nation's next generation of gentlemen leaders.

With this new dream in mind, Davis returned to Knoxville. However, he was unsure of how to raise funds and had to put this venture on hold. He continued working at his father's shoe business in the meantime, slowly trying to make connections with businessmen and investors in the local area. Davis joined the Democratic Party, which was a major source of power in the South, and became a member of the Royal Arcanum. Through these groups, Davis came into contact with local war legend Colonel Nathaniel Macon Taylor. However, it was one of the colonel's daughters, Bessie Taylor, who greatly interested Davis. The two began dating right away and got married on January 27, 1903. They moved to Bristol, Tennessee, where they made their home for the next few years. In 1906, they had their first child, Marinita.

The Military School Boom

While in Bristol, Davis took on teaching jobs and became involved with the local YMCA, spending most of his time teaching young students at the facilities. He eventually was promoted to president of the board of directors of the YMCA in 1907.

Around this time, countless military schools sprang up around the country, created by Spanish-American War veterans. These schools would pop up like daisies and then vanish as quickly as they appeared. It was a pattern common after major national wars, as a similar boom occurred right after the Civil War. However, most of the founders of these schools were not strictly intent on creating a strong educational institution; for many, these were intended to be easy postwar retirement jobs. Such schools would almost never survive the deaths of their founders. While hundreds of such schools were created after the Spanish-American War, most would last no more than a few years before closing. In fact, the only major military schools to survive from this era are Massanutten Military Academy (1899), Fork Union Military Academy (1902), Riverside Military Academy (1906) and Hargrave Military Academy (1909).

Most of these schools were purely religious schools first, predominantly Roman Catholic and Episcopal, and run by churches and nuns. While they styled themselves as military schools, they were primarily focused on teaching the Bible first, not leadership education, and used the military structure as a means to control the students.

Seeing the wake of this military school explosion in the South, Davis once again considered his own vision for a proper leadership-structured institution. If he did not pursue his dream quickly, he would be swallowed up in the sea of self-titled military schools created by retired officers and churches. Davis knew that if he created a school in his current location, it would be indistinguishable from all the others. Military schools were already popping up around him in Tennessee, notable examples being Baylor Military School (1893–1971), Castle Heights Military Academy (1902–86), Columbia Military Academy (1905–79) and Tennessee Military Institute (1909–75). To survive, he would have to create his school in an area with no competition. Thus, after gaining the needed experience, Davis and his family left Tennessee and headed west for Texas.

THE EL PASO MILITARY INSTITUTE

Captain Thomas A. Davis arrived in El Paso in May 1907. El Paso was a small desert town with nothing but dust stretching for miles. There was the Military Reservation, Fort Bliss nearby, but aside from that, there was nothing much of interest. For Davis, this was perfect. By now, military schools were being established in Texas; however, El Paso was so isolated from major cities that citizens of that area were unable to send their sons to any of these military schools.

The moment Davis arrived in El Paso, he began introducing himself to the townsfolk. He told them of his intention to establish a high-grade military academy in their town as the first in a larger system of military schools that would stretch across the western United States. Right away,

Davis was ushered to the Chamber of Commerce and Committee on Public Improvements, where he pitched his plan. The committee took an interest right away, being won over by Davis's charisma and his gentlemanly personality. The committee was made up of powerful individuals who held great influence over El Paso, the business and professional elite of the era. They took to Davis's plan for the creation and maintenance of such a school and approved the establishment of such a school in July of that year. The city formed a company of citizens interested in educational matters to help with the creation of the new school, to be named the El Paso Military Institute.

On July 10, the El Paso Military Institute received its charter from the State of Texas, which allowed land east of Fort Bliss to be donated by the city to be used by the institute, including eighteen acres by "public-spirited citizen" August Mesiel. In November, four additional acres of land south of the institute were donated by Charles R. Morehead of the State National Bank of El Paso and Captain T.J. Bell. Davis donated his own purchased land to expand the campus to twenty-two acres.

The Trost and Trost firm of El Paso, under architect Henry C. Trost, was employed, and Otto P. Kroeger constructed the buildings. The first board of directors was elected, including Judge Beauregard Bryan as president, H.B. Stevens as treasurer and Colonel Lewis as secretary. Davis was to be both the superintendent and commandant of the Cadet Corps. In 1908, the main building was completed and ready for opening day.

On October 1, 1908, the El Paso Military Institute opened its doors to eighteen cadets. Davis, happy to see his dream finally realized, cheerfully greeted the first class of incoming students and their parents. These boys came from all over the Southwest, as the institute quickly gained recognition from just word of mouth alone. To the families of the Southwest, finally having their own military school in arms' reach was a wonderful convenience. However, when the parents finally met Davis in person, his personality won them over completely. Davis's southern gentleman demeanor added immeasurably to the institute's reputation. Parents enjoyed being around him, and mothers especially noted him as being a fine gentleman.

The El Paso Military Institute was a labor of love by the citizens of El Paso. However, Davis wanted to make sure that his own personal touch would make the school unique compared to all the other military schools of that period. He wanted to make sure that the school was accepting of all students, even foreign students from south of the border. He made sure that the school was nondenominational; however, there was daily

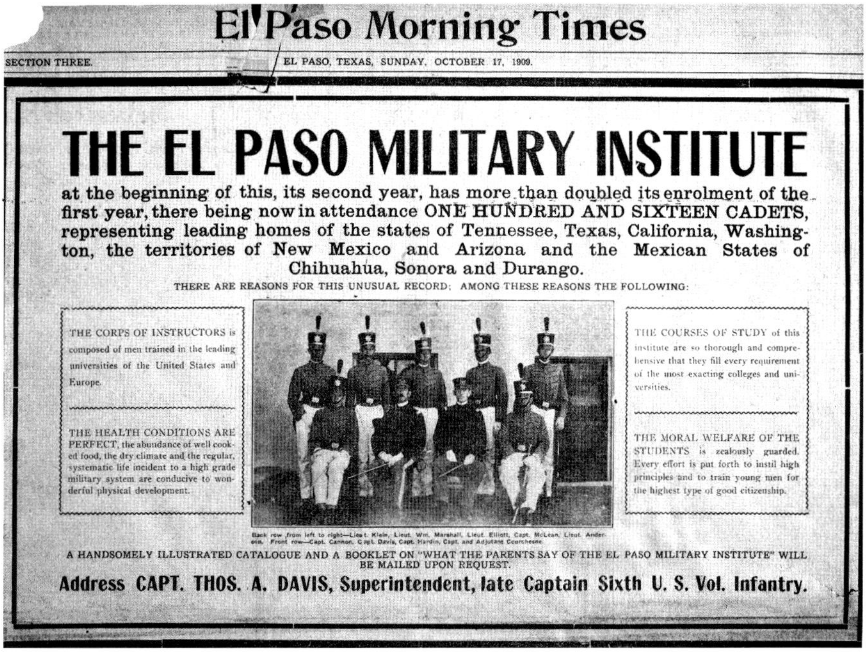

El Paso Morning Times

SECTION THREE. EL PASO, TEXAS, SUNDAY, OCTOBER 17, 1909.

THE EL PASO MILITARY INSTITUTE

at the beginning of this, its second year, has more than doubled its enrolment of the first year, there being now in attendance ONE HUNDRED AND SIXTEEN CADETS, representing leading homes of the states of Tennessee, Texas, California, Washington, the territories of New Mexico and Arizona and the Mexican States of Chihuahua, Sonora and Durango.

THERE ARE REASONS FOR THIS UNUSUAL RECORD; AMONG THESE REASONS THE FOLLOWING:

THE CORPS OF INSTRUCTORS is composed of men trained in the leading universities of the United States and Europe.

THE HEALTH CONDITIONS ARE PERFECT, the abundance of well cooked food, the dry climate and the regular, systematic life incident to a high grade military system are conducive to wonderful physical development.

Back row, from left to right—Lieut. Klein, Lieut. Wm. Marshall, Lieut. Elliott, Capt. McLean, Lieut. Anderson. Front row—Capt. Cannon, Capt. Davis, Capt. Hardin, Capt. and Adjutant Courchesne.

THE COURSES OF STUDY of this institute are so thorough and comprehensive that they fill every requirement of the most exacting colleges and universities.

THE MORAL WELFARE OF THE STUDENTS is zealously guarded. Every effort is put forth to instil high principles and to train young men for the highest type of good citizenship.

A HANDSOMELY ILLUSTRATED CATALOGUE AND A BOOKLET ON "WHAT THE PARENTS SAY OF THE EL PASO MILITARY INSTITUTE" WILL BE MAILED UPON REQUEST.

Address CAPT. THOS. A. DAVIS, Superintendent, late Captain Sixth U. S. Vol. Infantry.

Advertisement for the El Paso Military Institute in the October 17, 1909 Sunday copy of the *El Paso Morning Times*. *Author's collection.*

chapel aimed at instilling good morals. On Sunday, cadets were allowed to attend the churches of their choice. Davis stressed leadership education first and took to the development of an academic and character training program that became known as the Davis System. This systematic training—which balanced leadership, academics, ethics and manners—boosted the institute's reputation among the sea of military schools. Physical training, character building and individual instruction for each cadet were also stressed, along with military training. The institute quickly gained accreditation.

While Davis never attended West Point, he took the positive elements he saw from the United States Military Academy and integrated its structure into the new institute. West Point uniform was the only clothing allowed on campus. West Point–style education, rifle training, drill and military maneuvers were conducted on the parade grounds of Fort Bliss. Dress parades took place every Friday afternoon. Cadets were required to maintain good moral conduct and courteous behavior at all times and were not permitted to drink or use tobacco. The focus of the institute was to prepare boys who

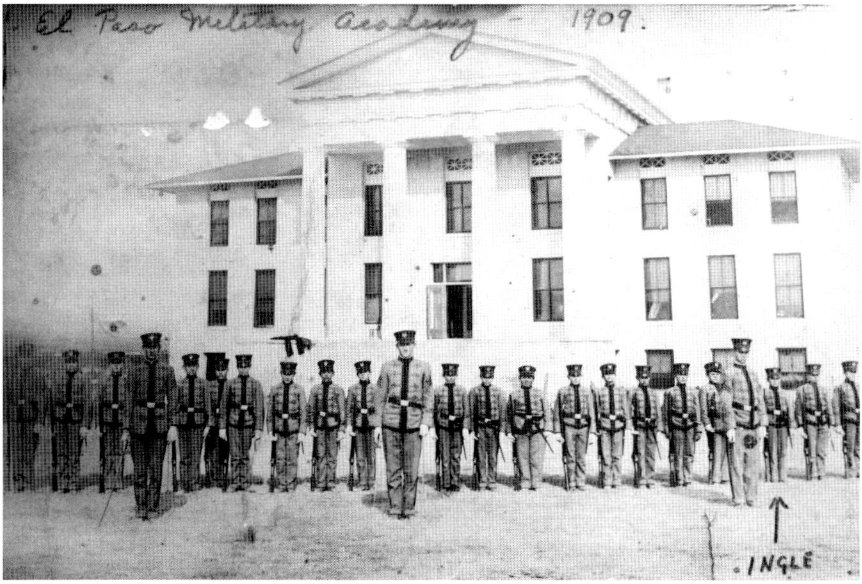

Photograph of the El Paso Military Institute in 1909. *Courtesy of Atkinson Museum.*

wanted to attend West Point or the Naval Academy, as well as any university of their choosing. And if a cadet chose, he could be commissioned into the U.S. Army upon graduation.

Admission to the institute required students to be of good character and ready to comply "cheerfully" with all regulations and requirements of life at EPMI. Boys had to be at least seven years old and in good health. Despite the age requirement, Davis did not sort cadets into class level based on age. He knew that each cadet, even those of the same age, varied greatly in ability, attainment and maturity. Therefore, each cadet was individually tested and placed in the classes and levels where he was best suited. As the enrollment size was manageable, courses of style and class selection could be customized based on the request of parents. There was a two-year primary class for younger cadets that would lead into the regular high school courses. Spanish was taught with great emphasis, and for international students, English classes were taught in a manner that would allow them to continue their regular classes.

The cost of admission for boarding cadets was $500 per year, while day students paid $100. This included tuition costs, use of equipment and membership to the local YMCA. Along with classes, the institute had many student organizations, including a literary society, a glee club, orchestra and marching band. All the musical groups were led by Major H.E. Van Surdam (1881–1982), who first joined the institute in 1909 as a science instructor and coach.

The institute's song was called "Sons of the South." The words to the original version were sung on Thanksgiving Day 1908 in Nashville, during the Sewanee-Vanderbilt game. The music and words were published by the Pass City Publishing Company and dedicated to the institute.

The El Paso Military Institute tripled enrollment within a year of its opening. Recognition for the institute grew throughout the Southwest. To cadets, however, the school was affectionately dubbed the "El Paso Mule Incubator."

In 1909, the first historical meeting between the presidents of the United States and Mexico took place in El Paso and Juárez. President William Howard Taft and President Porfirio Díaz met in both sister cities. Taft became the first American president to visit foreign soil. Both presidents paid the institute a visit. The cadets, who were one of the main attractions, lead the Taft-Díaz Parade in honor of the visit. Schoolchildren sang, and crowds shouted "Viva Taft!" and "Viva Díaz!" One cadet from Company C who marched in that widely received parade, Modesto Gomez, would later serve in the army and fight in the Meuse-Argonne Campaign in France during the First World War.

The regional recognition and visit from the sitting president within its first year would be more than enough for any military school. However, for Davis, he still felt that something was missing, that his vision had yet to be fulfilled. One can only assume that Davis felt trapped and isolated in the Southwest. He perhaps knew that if he stayed in El Paso, he would never get the chance of someday overseeing a much larger education system that spanned the nation.

In 1910, Davis attended an exhibit in El Paso, where a display sponsoring a San Diego firm caught his eye. Seeing this as a way out, Davis set his sights on California.

PART I

BY THE CITY'S NORTHERN BORDER

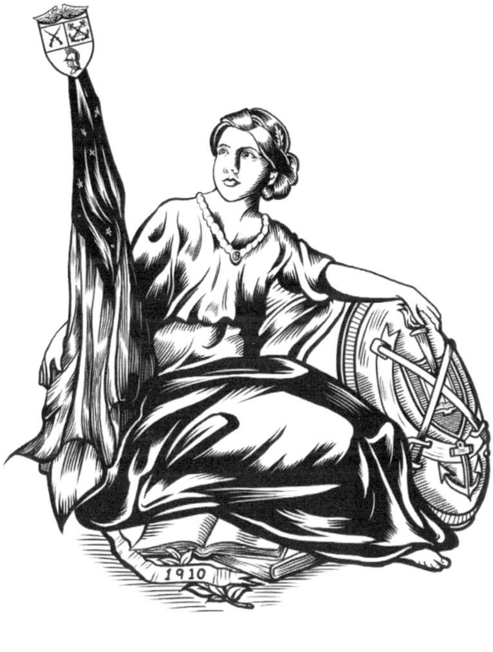

The Academy's Founding. Artwork by Alexander Mui.

CHAPTER 2

FOUNDING OF THE ACADEMY

IN THE SUMMER OF 1910, Captain Thomas A. Davis arrived in Pacific Beach, California, for the first time. This was the farthest west he had been, a continent away from his childhood home. Davis was instantly taken by the beauty of the area. "Attracted by the location of San Diego, its unequaled climatic conditions…I yielded to the request…to establish here a high grade military school," Davis would later remark. Little did he know at the time that this is where he would spend the remainder of his life.

There was no time for vacation, however. The instant Davis arrived at his new home, he set out with the intention to open his next military school within the same year. In June of that year, Davis became the sole leaseholder of the property that included the abandoned Hotel Balboa and twenty acres of land surrounding the old building. The Hotel Balboa had been a popular landmark for visitors traveling to San Diego for many years. It was located in City Park, which would be renamed Balboa Park in November of that year (now Pacific Beach Plaza). The initial area of land leased by Davis included all areas within Lamont Street, Ingraham Street, Diamond Street and Garnet Avenue. Despite the vacant area and its distance from other major buildings in Pacific Beach, Davis felt that the location between San Diego and Los Angeles was the perfect spot for his new flagship military school. He noted that the old hotel rooms were large, airy and elegantly furnished, perfect for new cadets.

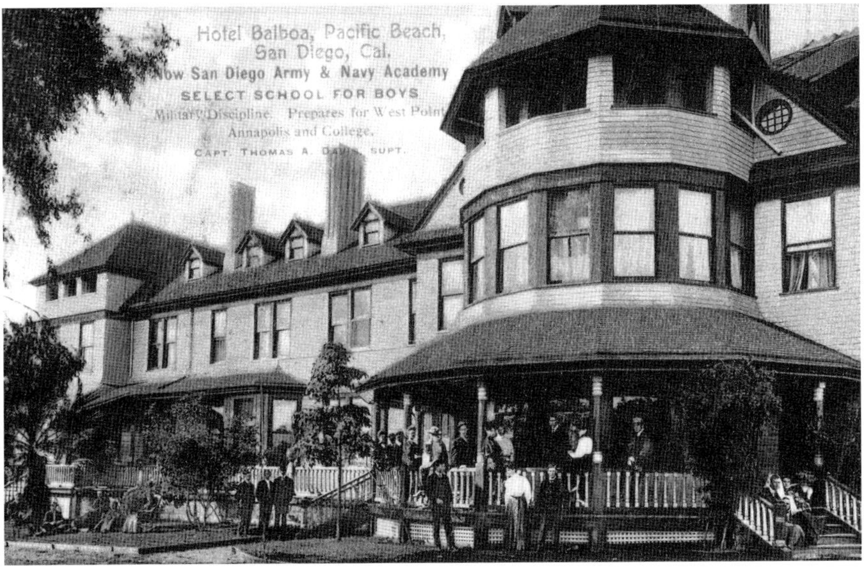

Postcard featuring the Hotel Balboa. *Author's collection.*

WESTWARD BOUND

Captain Davis left El Paso along with his wife and young daughter. Some parents who had sons at the El Paso Military Institute knew that Davis was the strength behind the school and felt that the institute would fail without him. They were eager to send their sons to Pacific Beach to continue at Davis's new school if it would open soon. In a race against time, Davis quickly went about trying to establish his new school before the waiting parents decided to send their sons to other military schools.

San Diego was the perfect location for Captain Davis's second academy. He noted that his choice of Pacific Beach was due to it being just nine miles from the business center of San Diego. The area was perfect for young men to get away from the distractions of the big city yet close enough for social visits and educational trips. He added that the area he chose was "especially fortunate as cadets can enjoy throughout almost every day in the year out-of-door classes, various athletic sports, in addition to the splendid boating and bathing on quiet Mission Bay, or the surf of the Pacific on one of the safest strands of the coast." Davis stressed that cadets would get to enjoy classes and formations in the fresh air and sunshine, which would have a more positive effect on their mental growth.

Davis traveled all across the San Diego area. His charisma instantly won over local communities. His personality brought him contracts with many local businesses and companies. The newly proposed school was originally called the San Diego Army Academy, but playing to the politics of the area, he added "Navy" to the name to gain support from the local San Diego Naval Base.

With the amount of work Davis had to do getting his new school up and running, he had no time to interview and hire faculty and staff. Unlike at the El Paso Military Institute, where a board was formed and many instructors were hired for the opening year, Davis intended to take up the burden of all positions for the initial year. Davis knew how he wanted to structure his school and knew that no one else could carry out his vision. Unlike at El Paso, with the Pacific Beach campus he wanted to remain in control over all areas of operations. Captain Davis would serve as sole instructor, Headmaster, Superintendent, Commandant of Cadets and professor of Military Science and Tactics for the first school year.

Founding Day

On November 23, 1910, the San Diego Army and Navy Academy opened her doors. Captain Davis personally greeted the first thirteen cadets and their parents. The opening day was also attended by many from the Greater San Diego area. This was seen as a major event for the community. The new Academy was not just the first military school to open in the San Diego area but also the first college preparatory military school opened in California.[2] Davis welcomed a large number of political and business representatives, including Secretary F.D. Fagg, Chairman Lorimer of the San Diego YMCA

2. The San Diego Army and Navy Academy might also be considered the first successful military school of any kind to open in California. While a few military schools did open in the state before 1910, most were primarily religious schools or reform schools that failed shortly after their founding. None was a college preparatory school like Army and Navy Academy. Records show that the first military school to open in the state might have been the Los Angeles Military Academy, which was founded in the 1890s, although it might have closed soon afterwards, as records of its existence are scarce aside from promotional material advertising its future grand opening. Other schools that might also lay claim to the title became military schools after 1910—for example, the Palo Alto Military Academy was founded as Manazanita Hall in 1893 but only became a military school in 1919, while St. Catherine's Military Academy was founded as an all-girls religious school in 1889 and only rebranded itself as a military school in the 1920s.

and citizens from all over Pacific Beach and San Diego. The invocation was given by Reverend J.T. French. Davis later recalled this event in *The Cadet* (1913), stating:

> *So on November 23rd, 1910, the opening exercises of the San Diego Army and Navy Academy were held in the auditorium of the large buildings which had been known previously as the Hotel Balboa, located in the center of Pacific Beach, the delightful suburb of San Diego.*

The members of the first class included W.D. Cannon Jr., William Fox, Harlow Garetson, Willard B. Hage, Sterling Hebbard, Forrest Henry Hieatt, Rex Minton Hocker, Charles Kingsley Johnson, James Hurvey Johnson, James Parker, Stewart McKee, Guy Parmenter and W. Armstrong Shaw.[3] Captain Davis dubbed this first Corps of Cadets the "Immortal Thirteen," which, he noted, "in this case [was] a lucky number, as I had only expected ten, because of the lateness of the date of opening."

A 1911 newspaper advertisement titled "A Select School for Manly Boys" stated:

> *Has been established at Pacific Beach in the elegant property formerly occupied by The Hotel Balboa. This is a preparatory school, under a military system, for boarding and for day students, which fits its cadets for practical life, for college, for West Point or Annapolis. The military drill gives wonderful physical development; strict military discipline produces habits of neatness, promptness and instills ideas of personal responsibility of honor, of manliness, and of patriotic citizenship.*

While Captain Davis was eager to expand enrollment size, he was not willing to dilute the reputation of the new Academy. The Academy started strong on a positive note and had made a name for herself in the local community; any student who might ruin that image would not be

3. The story of the Army and Navy Academy's founding day is often mixed. I have heard and read numerous accounts, including that the Academy was founded on November 10, 1910—this is perhaps conflated with the date of the United States Marine Corps' birthday. I also read that the first thirteen cadets were all day students, while another account states that the first cadets included thirteen boarding students and five day students. The Academy's official stance is that the founding day is November 23, 1910, with thirteen students of unstated status. The earliest physical account of the founding day I can find is from the 1913 yearbook, the story I recount here.

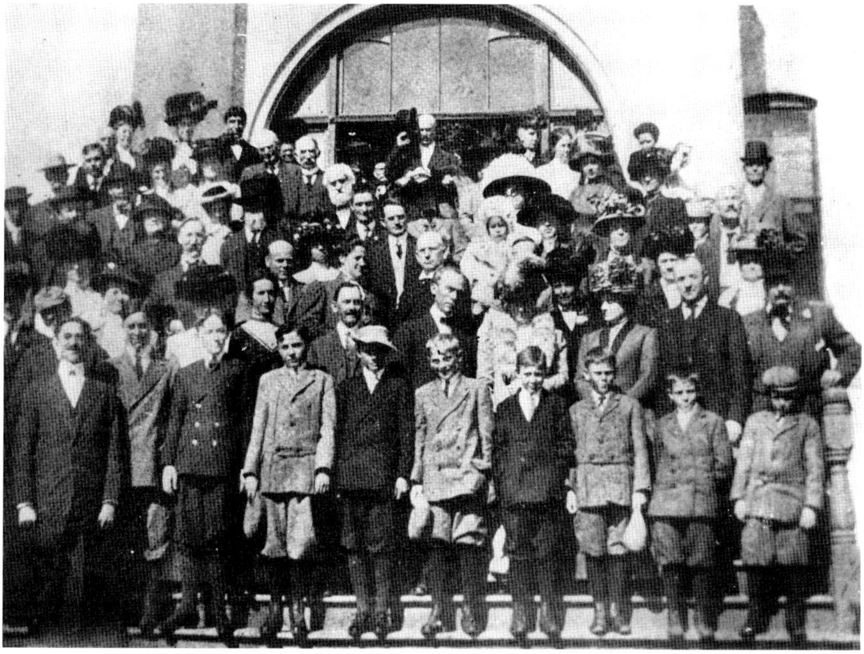

Opening day at the San Diego Army and Navy Academy, November 23, 1910. *Courtesy of Army and Navy Academy.*

welcomed. Davis chose only the best of applicants, as the Academy is a "select school for manly boys." Cadets need to stand for "clean speech, clean athletics, and clean life." Davis mentioned that if a parent were looking for a disciplinary reform school, he or she should look elsewhere.

An application submitted by a parent had to be accompanied by report cards and class records to ensure that prospective cadets were of good character. Under no circumstances were the smoking of tobacco or drinking of "spirituous or intoxicating liquors" allowed. Any cadet caught smoking or drinking was debarred from any promotion, and cadet officers lost their ranks for the remainder of the school year. He warned that repeated violations may result in confinement to the guardhouse or dismissal from the Academy. Another point Davis strongly stressed, and put in heavy italics, made his stance clear on certain practices: "*Hazing is positively forbidden, and cadets guilty of this offense will be dealt with summarily.*" While the hazing culture was commonplace in the military school community at this time, Davis stressed that such actions would not be tolerated and would result in automatic dismissal with no refund.

An early brochure for the Academy noted:

> *The Academy has no desire to enroll undesirable boys and it is presumed that a parent or guardian in making application for a boy, thereby vouches for his good character and represents him to be amenable to discipline and agrees that in the event of dismissal, suspension, expulsion, or for absence, or withdrawal, no deduction or refund shall be made from the charges of the full year and any amounts not paid shall immediately become due and payable as liquidating damages. This forfeiture by no means compensates for harm done a school by the admission and removal of an incorrigible boy.*

THE FIRST YEARS

The cadets of the San Diego Army and Navy Academy marched in their first parade during the groundbreaking day of the San Diego Exposition. Captain Davis led those thirteen cadets down Main Street, along with a Boy Scout who was borrowed for the occasion to carry a banner for the Academy. After the holidays, the Cadet Corps grew when it welcomed eight more cadets: George Andrews, Gaylord S. Clark, Edward J. Falk, Baldwin Haston, Paul Johnson, John Paton, James S. Whitney and Tracy S. Whitney.

For the first academic year, the old Hotel Balboa housed the classrooms, mess hall and residences for the tiny cadet corps. This time, he had complete control over all aspects of the school. Captain Davis continued his Davis System, which he developed and perfected from his years of working with the YMCA and El Paso Military Institute. His motto to the parents was, "To develop and train young men in Christian character and prepare them for the duties of American citizenship."

Davis was something of a miracle worker. He enhanced the reputation of the tiny school. Within months of her founding, the popularity of the Academy had spread quickly throughout the San Diego area, and many parents began to enroll their sons. By the end of the 1910–11 school year, the popularity of SDANA had grown tremendously, and her reputation was cemented. Letters of appreciation started pouring in from grateful parents, who began recommending the Academy to others.

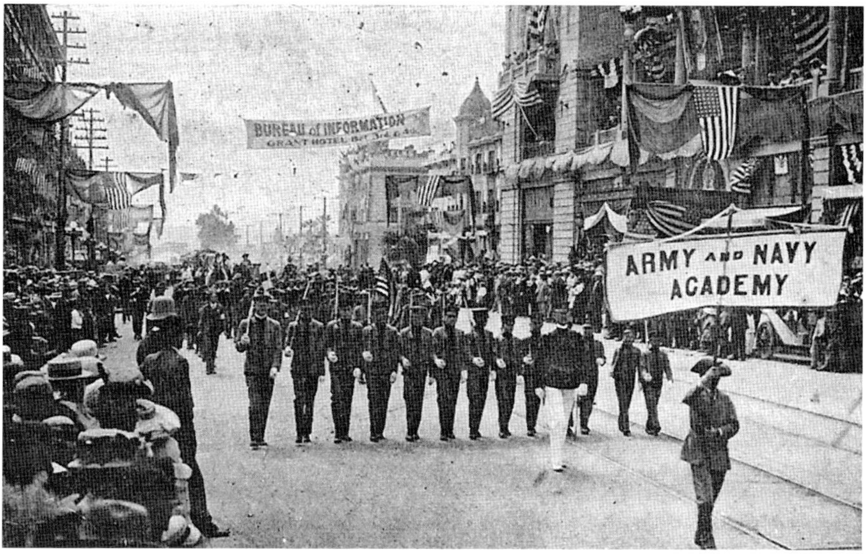

The San Diego Army and Navy Academy's first parade, at the groundbreaking day of the San Diego Exposition in 1911. A Boy Scout was borrowed to carry the banner. *Author's collection.*

As one parent wrote to Davis after the first term in a letter dated January 7, 1911:

> *We would like to let you know how we appreciate the way you have taken an interest in our son, since he has been placed in your Military School; we think it is one of the best schools that we have had any thing to do with, and our Boy has been benefited in a great many ways since he entered your school, and you surely understand how to work with Boys, and no one will make a mistake, in placing their boy in your school. Thanking you for what you have done for our Boy...*

As the public face of the Academy, Captain Davis was, as always, popular with the parents, who were impressed by the manners of this southern gentleman. On September 18, 1911, SDANA started a second year. By this point, the Corps had grown to fifty-eight cadets. Knowing the challenge of being the sole instructor the year prior, Davis decided to hire new faculty members. He noted that these were to be gentlemen of his choosing, meant to keep up with the growing Cadet Corps.

In 1911, the Academy's first Headmaster, Jesse A. Beadle, was hired. That school year, Davis brought on Ralph R. Rice as the modern language

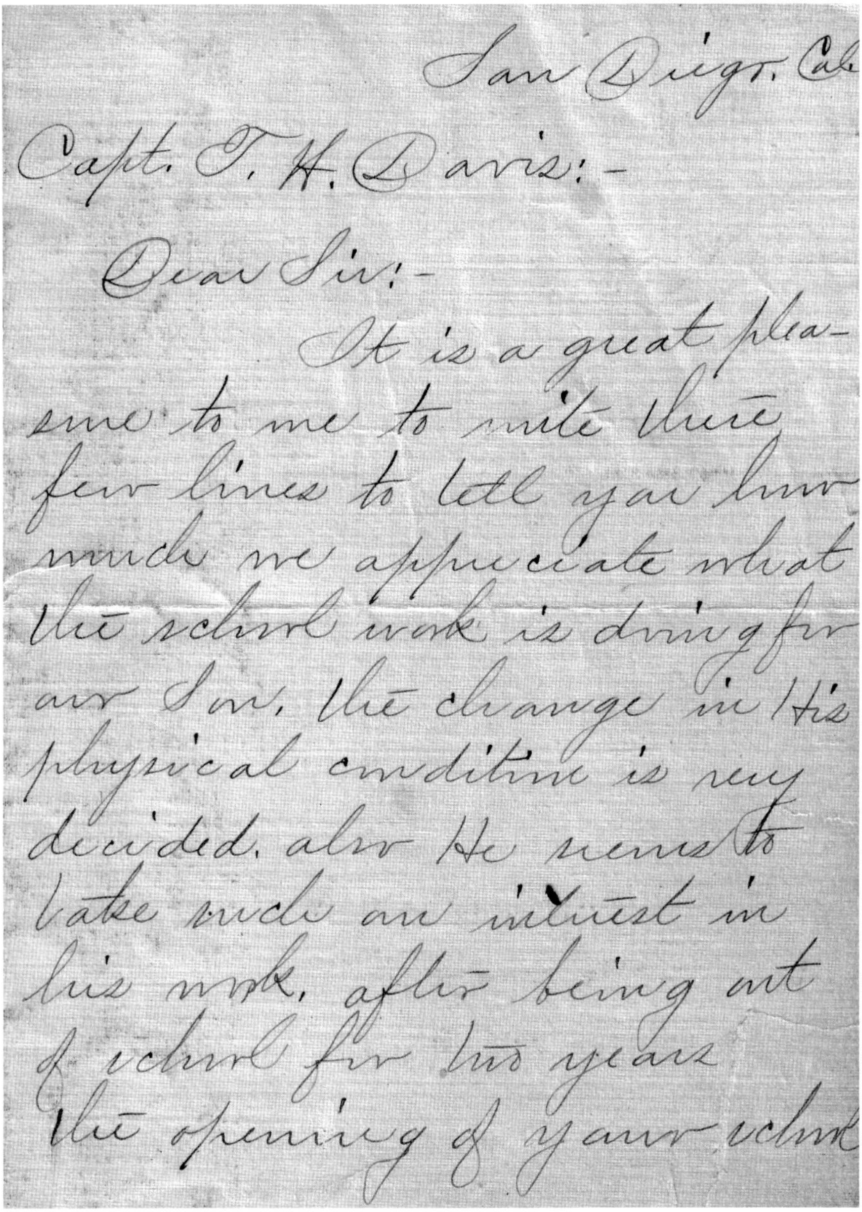

San Diego. Cal

Capt. T. H. Davis:-

Dear Sir:-

It is a great plea-
sure to me to write these
few lines to tell you how
much we appreciate what
the school work is doing for
our Son. the change in His
physical condition is very
decided. also He seems to
take such an interest in
his work. after being out
of school for two years
the opening of your school

A letter written by a satisfied parent thanking Captain Davis. *Courtesy of Atkinson Museum.*

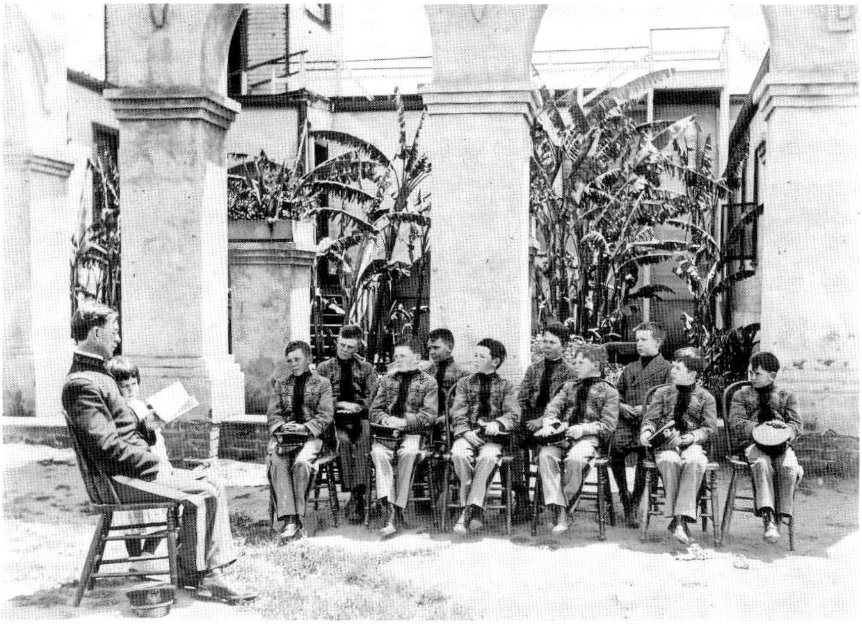

Captain Davis teaching cadets with his daughter seated near him. Outdoor classes were something Davis thought would benefit a growing mind. *Author's collection.*

instructor and C. Nelson Richardson as the financial administrator. Richardson's wife became the house mother, or matron, to cadets. Davis's wife remained in charge of the Home Department.

In early 1912, Davis brought on Charles Milton Wood, who got his law degree from DePauw University in Indiana, to serve as the history instructor. For the following school year, Reverend Newell Scott Fiscus taught English, Herman J. Baade taught mathematics and science, Chesley Mills from the San Diego Symphony Orchestra taught music and telegraphy and David Cole Wilson taught Latin. Wilson also served as the Academy's first director of athletics and head coach. Davis also hired Kate E. Breckenridge as his personal secretary and instructor of writing and typewriting courses to cadets. On selecting his faculty, Davis commented:

> *I believe that a teacher should take up work with boys only through a deep love for boys and with a holy purpose to direct them through the critical period of growth to the highest standards and ideals of manhood he must not think of a boy as a piece of wood or metal or as an insensible plant, but as one of God's creatures, palpitating with life and energy and brought into the world*

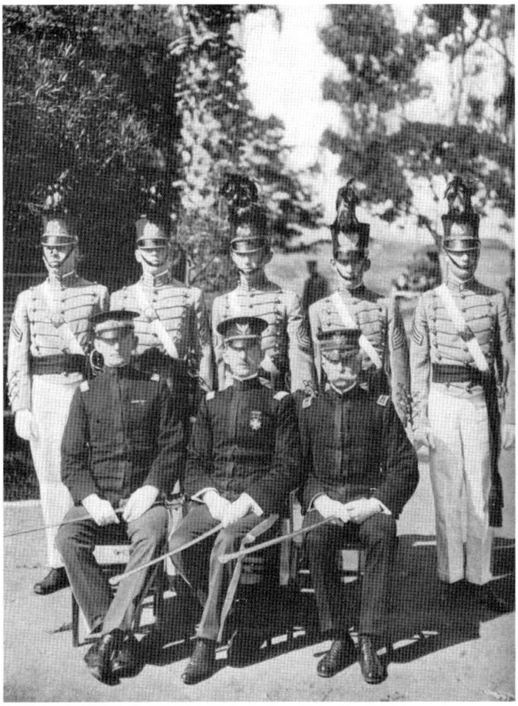

Left: Staff and cadets at the San Diego Army and Navy Academy. Captain Davis is seated in the front center, with Headmaster Beadle on the right. *Courtesy of Atkinson Museum.*

Below: Cadets and Davis's daughter seated for a photograph. *Author's collection.*

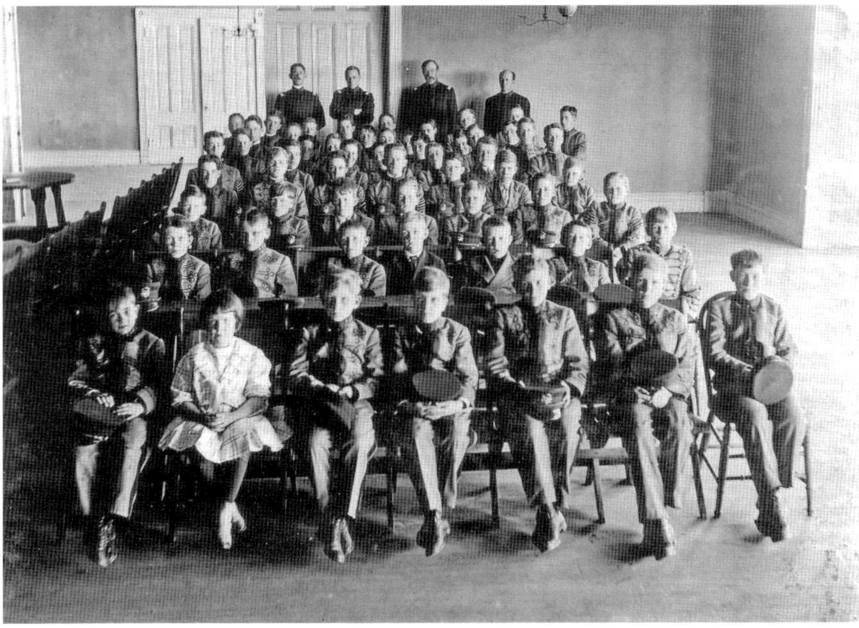

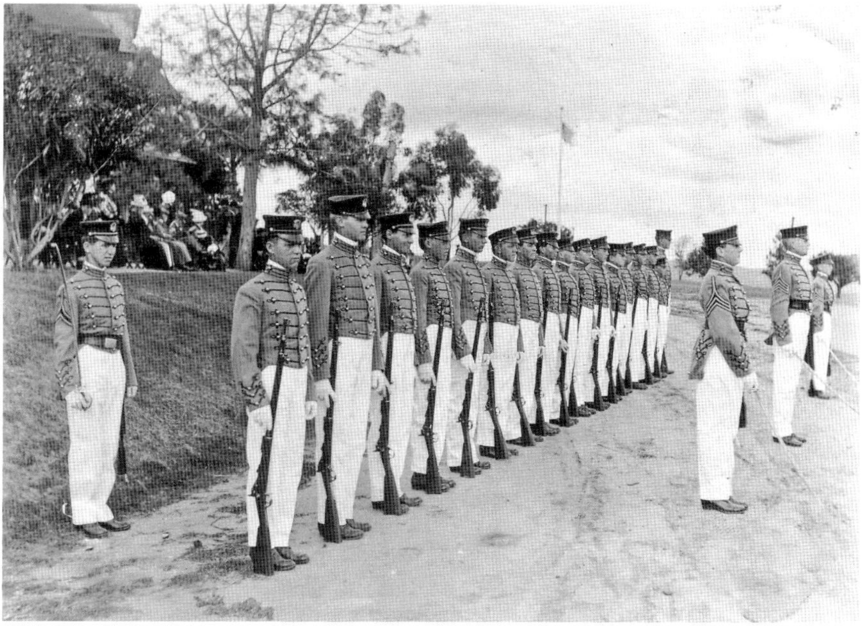

Cadets of the San Diego Army and Navy Academy in formation during a dress parade in Pacific Beach. *Author's collection.*

> *with varying impulses and special natural endowments. A teacher should not only be an instructor, but also a friend, a helper, a comrade, a big brother to the boy under his care and it has been my constant aim.*

The monthly cadet-run school paper, *The Cadet*, came into existence that year. It would become a tremendous influence on the lives of the Corps for years to come. At the end of the year, *The Cadet* would publish an annual special edition as a keepsake for everyone at the Academy. This would eventually evolve into the Academy's yearbook.[4] By the second year, the Corps had multiplied more than tenfold. Charles Wood exclaimed, "But a jump from thirteen pupils during the first fractional year to one hundred and forty in the second full year is really not bad jumping, and while speaking well for the popularity of the school, and for the work being done, it also augurs well, if you please, for larger graduating classes in a few years."

4. *The Cadet* served a dual propose, the first section making up an annual yearbook and the last section an advertising catalogue for the school. The cadet newspaper was also called *The Cadet*.

THE FIRST WARRIOR TEAM

Sports are not only a source of pride but also a large part of the history of any educational institution. This is no exception for the Army and Navy Academy Warriors. While the Academy was initially too small for a sports team, when enough cadets showed interest, Captain Davis allowed for the formation of small teams.[5] In the first year, the Academy established her first football, basketball, baseball, tennis, track and swimming teams. "No climate in the world would allow out-of-door sports throughout so much of the year as does the climate of San Diego," Davis noted. However, he stressed, "Athletic features, while encouraged, are not allowed to interfere, however, with any cadet's progress in his studies."

The General Athletic Association (GAA) was formed in the later part of October 1912, organized by Headmaster Beade and Head Coach Wilson, with cadets Harlow Garetson as the club's president and Thomas Scripps as treasurer. This group was the forerunner of the Academy's Athletic Department and was meant to serve as the regulating body for all campus sports teams. Any cadet who wanted to play sports needed to be a part of the GAA, which also raised money to pay for uniforms and equipment. Initially, the club did not draw enough interest. The 1912–13 yearbook mentioned, "As a responsible body the GAA has been a failure, because of the lack of interest shown by the student body." It was mentioned that Davis eventually had to put up the money to financially support the club and all the athletic teams; the following year, the club proposed an athletic fee of ten dollars for all cadets to begin the next academic year financially strong.

That same year, SDANA formed her first competitive football team. Athletic Director Wilson served as the head coach. Cadet George Bean, the halfback, was the first team captain. Bean was noted as "one of our stars, and certainly deserves much praise for his work in the back field and as captain of the team." On October 1, a group of thirty-five cadets turned out for tryouts. After two weeks of heavy practice, fifteen players were whittled down to form the first football team.

The Academy entered the Southern California Football Conference and played her first game against Santa Monica, resulting in the team's first defeat. Because the football team and practice started late in the season, the Academy was at a disadvantage and was swiftly defeated in her first games against Escondido, San Diego High and Pasadena. The young team was not

5. Early records (as noted in the 1912 and 1913 editions of *The Cadet*) reveal that Davis was somewhat opposed to organized "professional" sports on campus initially.

the strongest in the area, as that year's cadet newspaper put it—"Although a young school and older schools can beat us in football, suppose we have a competitive drill and see who holds the long end of the score?" Commenting on the Academy's first football season, the yearbook noted:

> *It may seem to many, by the number of games we lost, that it was an unsuccessful year, but far from it. It put us alongside of other preparatory schools larger than ourselves.... We played a clean season of football, and always gave all we had in us when we played. It must also be remembered that this was our first year of organized football.*

In the final game of the season, the Academy finally achieved her first victory, against South Pasadena. The team noted that the game "was ours from start to finish."

The first basketball team was formed in time for the winter season, in a promising start. The Academy's basketball team played its first game on January 21, 1913, against the Coast Artillery Guards. Cadets cheered the new basketball team on to its first victory.

CLOSING OF THE INSTITUTE

Without Davis, the El Paso Military Institute marched on. The Mexican Revolution erupted in 1910, leading to greatly reduced operations at the institute. Sadly, as the years passed, financial problems became too great an obstacle for the institute, and there was not enough money to continue supporting the school. In 1912, Major Surdam, who had joined the institute in 1909 as science instructor and coach, became the superintendent and served in this position for the remainder of the institute's existence.

While it was evident that without Captain Davis, EPMI was doomed to fail, the institute's football team dominated in the area, playing high school teams, college teams and even military teams. During the 1911 season, the institute beat Fort Bliss (30–0) and the El Paso Tigers (27–0) during the El Paso City Championship Game. In the 1912 season, the institute beat the U.S. Army 22nd Infantry (13–6) and U.S. Army 13th Cavalry Troop M (9–3). The only area rival that dominated the institute was New Mexico State, which consistently won, in 1910 (0–12), 1911 (0–3) and 1912 (0–40).

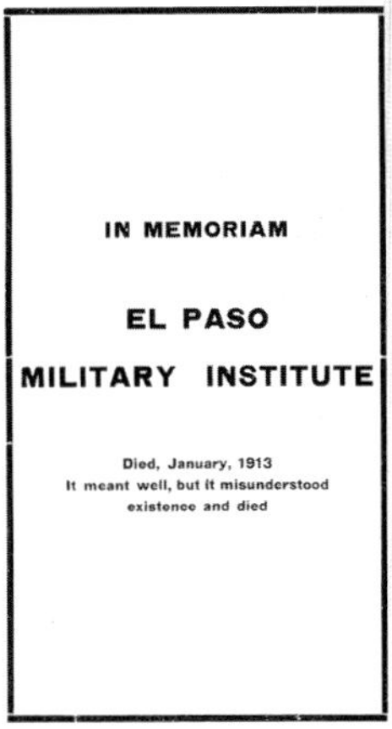
IN MEMORIAM

EL PASO
MILITARY INSTITUTE

Died, January, 1913
It meant well, but it misunderstood
existence and died

A humorous dedication to the El Paso Military Institute in the 1913 yearbook of a local El Paso school. *Author's collection.*

On November 21, 1912, Major Surdam met with Texas governor Oscar Colquitt and William B. Phillips, director of the University of Texas's Bureau of Economic Geology, to discuss the future of the EPMI. Seeing that the school would soon be forced to close, the cadets were determined to finally beat their rivals, New Mexico State.

On November 23, the very day SDANA celebrated her second anniversary, Captain Davis's first military school played rival New Mexico State for its final game. Up to this point, the young military high school had never pulled out a victory over the much larger university football team. Playing on home field for the last time, the EPMI football team soundly defeated New Mexico State (20–0), a bittersweet ending for the institute.

In January of the following year, the El Paso Military Institute was forced to close. The property was handed over to the State of Texas, which turned it over to William B. Phillips and the University of Texas. The institute was converted into the Texas School of Mines and Metallurgy for the University of Texas, which began classes in September 1914. On October 29, 1916, a fire destroyed most of the buildings on the campus. Fort Bliss bought the property back and turned the surviving El Paso Military Institute dormitory into the Fort Bliss Airdrome. The School of Mines had to move to a new location within El Paso. In 1919, the school was renamed the College of Mines and Metallurgy. Its name was later changed to Texas Western College before becoming the current University of Texas–El Paso.

Although it had a short life, the institute had a strong influence on its cadets, many of whom grew up to be prominent citizens in El Paso and throughout the Southwest. Many fought in the Great War. One notable cadet was James B. "Buster" Biggs, who served as a lieutenant in the United States Army Air Corps. In 1919, he was killed in a flying accident in Europe. The Biggs Air Force Base in El Paso was named in his honor.

On November 18, 1932, the former cadets of the El Paso Military Institute gathered for a reunion for the first time and discussed creating an alumni association. For the next decade, only memories would remain of EPMI, which was considered among the best of military preparatory schools. With time, even those memories have faded away.

CHAPTER 3
THOSE DISTANT YEARS

The San Diego Army and Navy Academy opened with thirteen cadets. That year's football team had thirteen players. Some would note this to be an unlucky sign for Captain Thomas A. Davis's new military school. He probably reminded those suspicious few that America was founded with thirteen colonies. He would later say of this, "Thirteen to me is an interesting number. I opened my school with thirteen. And three years later, or in 1913, I had a multiple of 13 occur several times in connection with the school." The year 1913 was indeed a notable one for the young school, as that year the University of California officially accredited SDANA.

That year, perhaps to celebrate the accreditation, Captain Davis surprised the cadets with a boat trip. Davis charted the *Empress*, a glass-bottomed boat, and took the entire school for an afternoon and moonlight ride around San Diego Bay. They then were met by Admiral W.W.H. Southerland, who invited the cadets aboard his cruiser, *Maryland*. The cadets then sailed to Fort Rosecrans and got to tour the batteries before returning to Santa Fe Wharf.

A YOUNG SCHOOL

A convenient car service created between Pacific Beach and San Diego aided the San Diego Army and Navy Academy's growing enrollment. The Academy was also accessible to out-of-town folks via the San Diego Beach Railway, as known as the La Jolla Line.

In 1912, the cadet population grew to 156, more than ten times the amount the Academy opened with. With a Cadet Corps of that size, it was now possible to organize four cadet companies and a cadet band, which doubled as an orchestra. The faculty expanded to fourteen men and women, all personally selected by Captain Davis. Davis mentioned the Academy's rapid growth in a 1912 newspaper advertisement:

> *To take care of this unusual growth, it has been necessary to secure, in addition to the original buildings, five cottages for the use of the members of the faculty and seven tent houses for the use of other employed help and for recitation purposes. I believe that such a growth would have been impossible if the work of the Academy had not been thorough and if parents had not believed in the high ideals which have been striven for.*

Along with the five new cottages and seven tent houses, the twenty-acre campus saw the construction of new tennis and basketball courts and a baseball field. Two large buildings were used as the main facilities on campus, one being used for dormitories and the other for the recitation rooms, laboratories, assembly hall and the Boarding Department. Davis made sure that every room in these buildings was welcoming, with large windows that allowed plenty of sunshine. If the weather was pleasant, and it normally was year round, Davis would also allow for classes to be held outdoors. Cadets would pull up chairs and learn from their instructor while embracing the fresh air and lovely outdoor scenery.

That school year, the Academy's band played at the dedication of the new YMCA building in San Diego and at the Spreckel's Theatre Flood Benefit to raise money for the victims of the recent disaster. In May, the Corps of Cadets surprised the Editors of California organization, which was holding its annual convention in San Diego. The band, in full dress, formed at the station and played for them as they got off the train. The curiosity of the editors caused them to hold off attending the convention to visit the campus, where they were treated to a full-dress parade followed by a mock battle on the beach.

The 1912–13 school year produced the Academy's second graduating class, made up of two members: Paul Lewis Singer, captain of Alpha Company and president of the Glee Club, and Herbert Chassee Tiffany, captain of Charlie Company and captain of the basketball team. Charles Wood, advisor to the senior class, remarked:

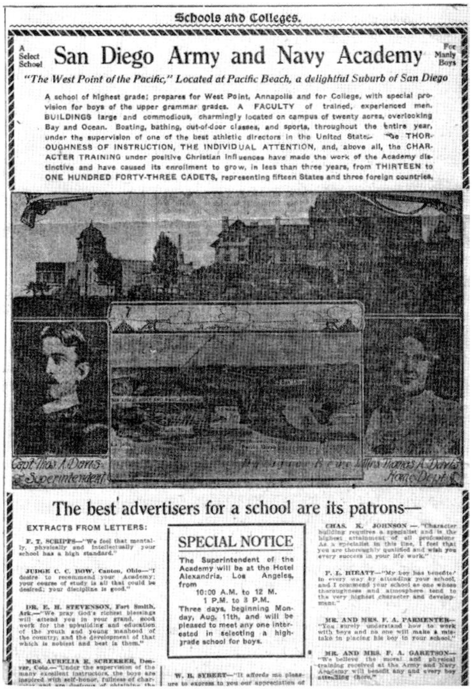

Schools and Colleges.

A Select School **San Diego Army and Navy Academy** For Manly Boys

"The West Point of the Pacific," Located at Pacific Beach, a delightful Suburb of San Diego

A school of highest grade; prepares for West Point, Annapolis and for College, with special provision for boys of the upper grammar grades. A FACULTY of trained, experienced men. BUILDINGS large and commodious, charmingly located on campus of twenty acres, overlooking Bay and Ocean. Boating, bathing, out-of-door classes, and sports, throughout the entire year, under the supervision of one of the best athletic directors in the United States. The THOROUGHNESS OF INSTRUCTION, THE INDIVIDUAL ATTENTION, and, above all, the CHARACTER TRAINING under positive Christian influences have made the work of the Academy distinctive and have caused its enrollment to grow, in less than three years, from THIRTEEN to ONE HUNDRED FORTY-THREE CADETS, representing fifteen States and three foreign countries.

Capt. Thos. A. Davis Superintendent

Mrs. Thomas A. Davis Home Dept.

The best advertisers for a school are its patrons—

EXTRACTS FROM LETTERS:

F. T. SCRIPPS—"We feel that mentally, physically and intellectually your school has a high standard."

JUDGE C. C. BOW, Canton, Ohio—"I desire to recommend your Academy; your course of study is all that could be desired; your discipline is good."

DR. E. H. STEVENSON, Fort Smith, Ark.—"We pray God's richest blessings will attend you in your grand, good work for the upbuilding and education of the youth and young manhood of the country; and the development of that which is noblest and best in them."

MRS. AURELIA E. SCHERRER, Denver, Colo.—"Under the supervision of the many excellent instructors, the boys are inspired with self-honor, fullness of char-

SPECIAL NOTICE

The Superintendent of the Academy will be at the Hotel Alexandria, Los Angeles, from
10:00 A.M. to 12 M.
1 P.M. to 3 P.M.
Three days, beginning Monday, Aug. 11th, and will be pleased to meet any one interested in selecting a high-grade school for boys.

W. B. SYBERT—"It affords me pleasure to express to you our appreciation of

CHAS. K. JOHNSON—"Character building requires a specialist and is the highest attainment of all professions. As a specialist in this line, I feel that you are thoroughly qualified and wish you every success in your life work."

F. L. HIEATT—"My boy has benefited in every way by attending your school, and I commend your school as one whose thoroughness and atmosphere tend to the very highest character and development."

MR. AND MRS. F. A. PARMENTER—"You surely understand how to work with boys and no one will make a mistake in placing his boy in your school."

MR. AND MRS. F. A. GARETSON—"We believe the moral and physical training received at the Army and Navy Academy will benefit any and every boy attending there."

An advertisement for the Academy in the August 10, 1913 edition of the *Sunday Times. Author's collection.*

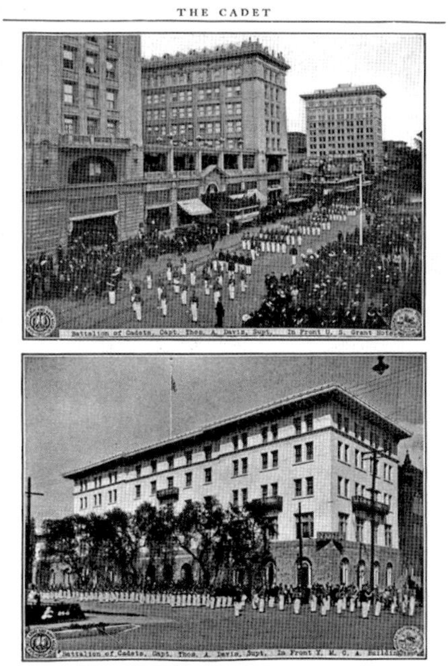

Top: The Cadets Corps marching in the San Diego Decoration Day Parade. *Bottom*: The Cadets Corps in front of the Army Navy YMCA. *Courtesy of Atkinson Museum.*

> *This year the Academy sends forth its second graduating class.*[6] *As formerly, the class is small in numbers, but we believe that what it lacks in numbers is more than compensated for in quality. Proud as we justly were of last year's class, that of this year ranks well with it, and we confidently look for as great and good things from our boys of 1913 as from any that have gone out from us hitherto, or that may hereafter go.*

That year, the Academy became affiliated with the University of Southern California, Pomona College, Occidental College, University of Redlands and the University of Tennessee. In 1913, the University of California officially recognized the Academy and gave accreditation with its highest rating, Division A, an honor shared by only two other military schools. Captain Davis recalled in that term's yearbook, "When the examiner from the University of California was shown the plans used by the Academy, whereby the grading and general record of each boy are kept under the observation of every member of the faculty, and the methods for securing effective work were explained, he said, 'In my ten years' experience, I have not seen a better system for looking after the individual student.'"

Davis mentioned that if any prospective parent wanted to learn more about his school, he would cheerfully send them to any bankers or businessmen in San Diego or to the examiners of the various colleges and universities, all of whom would vouch to the quality of the Academy. In the 1913 catalogue for the Academy, Davis wrote:

> *The number in each class has been limited so that individual help and attention could be given; thus the interest and enthusiasm of the pupils have been aroused to such an extent as to cause an examiner from one of the large colleges to remark on a recent visit to the Academy, "There is a different atmosphere around this school; one notices it immediately on visiting your classes; these boys are alive."*

The variety of classes also grew as more cadets were enrolled. Along with the basic core classes, military maneuvers and use of firearms, other courses included chemistry, geology, mineralogy, physics, Latin, Spanish, French, dramatic art, music and athletics. Davis would make sure that the cadets were given the best care and taught by the best of teachers. He made sure all

6. Based on this detail, it is most likely that the Class of 1912 was the Academy's first graduating class. Currently, it is not known who made up the first graduating class or how big it was.

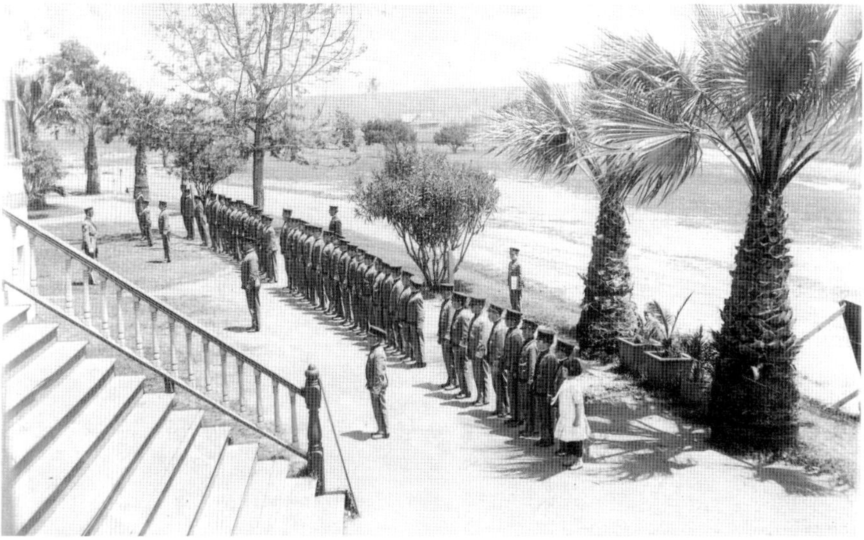

Cadets in formation outside the main building. Davis's daughter is seen at the end of the formation. *Author's collection.*

the cadets would fulfill their duties as good Christians and Americans. Most of the teachers he hired on his small staff were devoted Christians who had served in the military.

During the 1913–14 school year, Europe entered the Great War, and although the United States would not be involved for a few more years, the effects of the war were instantly felt. The world was changing—the era of colonization and imperialism was now giving way to an era of global communication and conflict. The Great War marked the beginning of a new period for the Academy. The faculty and staff slightly dropped as the war began, but Captain Davis made the effort to maintain the quality of education. Despite the setback, the efficiency and spirit throughout the Corps remained extremely high.

One member of the Class of 1914 attended the University of California, being the first Academy cadet to attend and graduate from the UC system. This would mark the beginning of a new relationship between the Army and Navy Academy and the University of California, as for the next century countless cadets from the Academy would continue on to UC schools.

THE WEST POINT OF THE WEST

In the fall of 1914, the Academy secured an army officer to conduct the Military Science and Tactics courses under the "Notice of Government Regulation" from the War Department. The United States Army added the officer based on the success and reputation of the Academy.

That same school year, on April 7, 1915, the San Diego Army and Navy Academy was officially accredited by the United States Military Academy at West Point under the provisions of General Order No. 19. SDANA became the first Pacific Coast private school to attain West Point accreditation, and this later forged a unique relationship between these two campuses. This relationship allowed cadets from the Pacific Beach campus to forgo many of the primary tests given to most other applications, including bypassing the mental examination, among others. Cadets who enrolled with intention of attending West Point or Annapolis received special arrangement and classes designed for acceptance to the federal academies. Later, at least one seat for the incoming class at West Point and other service academies was guaranteed to a cadet from the Army and Navy Academy, which had the ability to nominate one cadet to each service academy every year. Since then, at least one graduate from the Army and Navy Academy has been accepted to West Point and Annapolis every year. As SDANA was the first West Coast school to gain West Point accreditation during this period, Captain Davis began calling his Academy the "West Point of the West."[7]

Captain Davis would continue to travel around the state to help the Academy gain recognition throughout California. As one advertisement read, "The individual care and attention, the thoroughness of instruction, and the character building have made the Academy distinctive." Local publications described him as a traveling medicine man whose charm and charisma would not just convince parents but also persuade young men, who were eager to enroll after his speeches. Everywhere he went, whether for business or recreation, he would promote the Academy.

As the Academy's fame grew, prospective students from distant locations started enrolling. Cadets were reported as coming from all across California, Minnesota, Colorado, New Mexico and fifteen other states, as well as British Columbia, France, Canada, Mexico, Latin America, the

7. While other schools have used the nickname the "West Point of the West," the San Diego Army and Navy Academy had the first recorded use of this title. This nickname, in most cases, refers to ANA in most publications. More details are contained in the Additional Source Notes.

South Pacific and China. In a progressive move, Davis welcomed many foreign students shortly after opening the Academy, being one of the first boarding schools of the nation to do so when most others still were restricted to white Christian citizens.

Around this time, price for admission was $600 per boarding student and $100 per day student. The summer session had a tuition of $100 for boarding and $30 for day students. Uniform cost, which included a West Point full dress, was about $50.

The average day for cadets started with reveille at 0630, followed by exercise, breaking and room cleaning. Inspection of quarters occurred at 0800, followed by classes until 1515. Then, from 1515 to 1800, cadets would partake in athletics or boating, followed by dinner. There was an hour of recreation and then a two-hour study period under the supervision of an instructor. Taps sounded at 2130.

With a growing enrollment, Davis was forced to find additional facilities. At this point, the campus consisted of two buildings—one was used as a dormitory and the second as an auditorium, dining room, kitchen, study

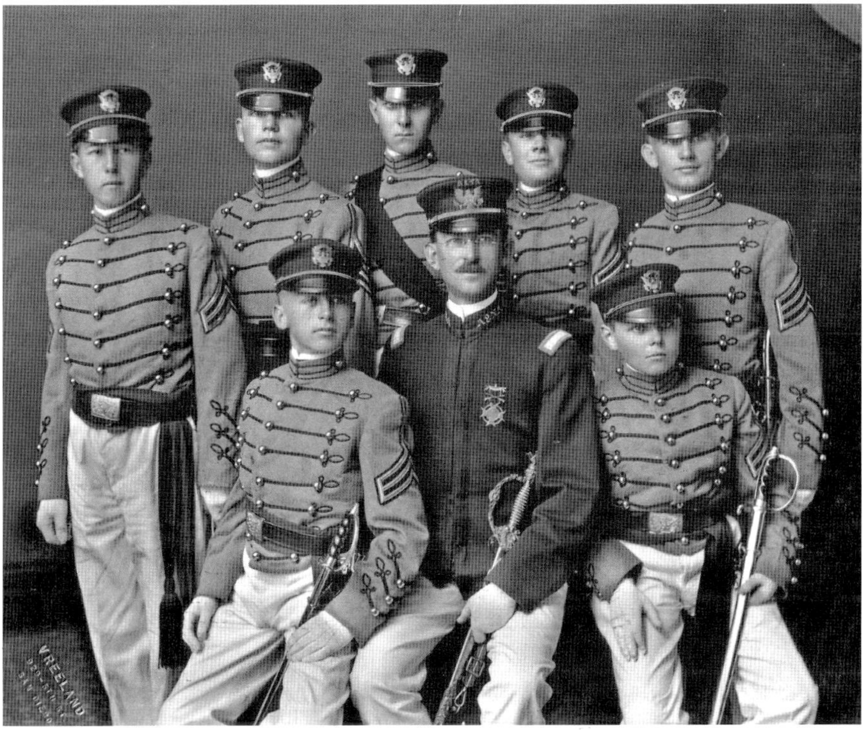

Captain Davis with cadets. *Author's collection.*

quarters, laboratories and classrooms. Davis bought two private homes for quartering and added eight temporary structures to accommodate the growing number of boarding cadets. He would also offer his home for the teachers who had no residence in the local area. However, Davis still had all the property on a loan and needed to find a means to pay them off.

In 1915, due to the tremendous cost associated with the growing enrollment, Davis was forced to organize a stock company. He said that the purpose was to take over active financial management of the Academy. Davis acquired the necessary special license to sell stock. The initial authorized capitalization was for $300,000. The members of this company included giants of San Diego's business elite, such as Ulysses S. Grant Jr., Fred T. Scripps of Scripps-Howard media giant, John D. Spreckels of Spreckels Sugar conglomerate and numerous bank presidents, admirals and generals, as well as prominent names from the legal, medical, real estate, merchandising and stock and bond fields.

In 1915, the Academy opened her first library and formed her first alumni association. SDANA was now fully accredited by the State of California. The War Department in Washington, D.C., gave the school a rating of M, an Essential Military School. In the first half decade of the Academy's existence, she gained more recognitions and honors than most schools do in their lifetimes. A notable member from the Class of 1915 was Fred D. Fagg Jr., who was president of the University of Southern California from 1947 to 1957 and chief of the Air Commerce Bureau with the United States Department of Commerce.

THE GREAT WAR

The new military academy came under the influence of her first war as the conflict in Europe finally hit America in 1917. With the United States joining the Great War, all plans for Academy expansion came to a halt. The graduated cadets who were educated at SDANA would finally face their first great challenge and put their Davis System training to use. Davis knew that this was exactly the type of war he was training his cadets for and prayed that he had taught them well.

All building materials at the Academy originally slated to be used to build new dorms were sent to the military for the construction of weapons and tanks. Many of the instructors and teachers also went off to fight in Europe.

Nevertheless, during these years the Academy gained three prominent faculty members. One was Major Samuel Warfield Peterson, who joined on as Headmaster, and another was James Allaben, who taught English and history. The third was Captain Davis's younger brother, Lieutenant John Davis Jr., who joined in 1917 as assistant professor of Military Science; however, he shortly left to join the war effort. Only nine seniors graduated that year, and just two seniors graduated the next. During the final year of the Great War, the influenza pandemic hit the eastern seaboard. The Academy was forced to close for two months; nonetheless, when the Academy returned to normal operations during the second semester, the examiner of schools awarded the Academy its highest scholastic rating. The federal inspector recorder also gave the Academy a very high rating.

The Class of 1917 had its Junior-Senior Banquet on June 1 of that year in the Academy dining room. This was an event held by Captain and Mrs. Davis each year for the graduating senior class and the upcoming senior class. This is the oldest recorded instance of the Academy's Junior-Senior Banquet tradition. During the dinner, the cadets wore orange and black, the colors of their graduation class.[8]

The senior class then attended the San Diego Bishop's School Senior dance the following day. June 3 saw the baccalaureate sermon for the graduating class, offered by Reverend C. Raymond Gray at Park Place Methodist Church. On June 6, the senior class had a reception and ball, the Senior Promenade, at the Club Room of the Pacific Beach Reading Club. This event was filled with 150 guests and decorated in orange and black, with walls draped with streamers and a large lantern of those colors. Charles Scott, who wrote one of the Academy's first songs, presided over the punch bowl at this event. The Class of 1917 graduated with a commencement ceremony on June 8. The yearbook jokingly added that there were no senior fatalities during this event.

After the War

By the time the Great War finally ended, the world had been greatly changed. The war that began with charging horses and sabers ended with tanks and

8. Based on this mention in the 1917 yearbook, it is assumed that during this time each graduating class had its own unique class colors or the graduation class was represented by the colors orange and black.

aerial fights, rewriting everything about mass warfare. It was then that the nation realized that on the battlefield, it was not brute strength or numbers that were of relevance but rather the need for men with well-rounded minds, moral judgment and strong education in leadership.

As the Academy emerged out of her first war, she received countless words of admiration from across the nation. During the Great War, seventy-two former cadets from SDANA served the United States overseas in military combat. Out of the seventy-two cadets who served, twenty were commissioned officers.

After being honorably discharged, Lieutenant John Lynch Davis Jr. returned from Europe to rejoin his brother at the Academy. By now, the Davis family had become fully involved with the operations of the school. Bessie Davis, the captain's wife, who was noted in publications at that time to be a very attractive and deeply religious lady, had taught Bible classes and headed the Home Department since opening day. Their daughter, Marinita, who followed her father to Pacific Beach when she was just a little girl, was then a graduate from Stanford University and became the social secretary for the Academy. His youngest child, Charles Andrew Murray Davis, was still a little boy at this stage, born after the move to Pacific Beach. He would join the family operations at the Academy for a while but later leave and go into a very successful business. Captain Davis's parents also joined the Academy. John Lynch Davis Sr. served as business manager and was noted for being very thrifty in his job.

During the 1919–20 school year, Lieutenant Davis formed the "Camel Club" to improve military bearing and physical strength, and it proved very successful. The same year, the Academy developed special rewards to recognize the hardworking cadets. The "Orderly System" honored a cadet each day for outstanding effort in some field of military, athletics or study. The "Purple Ribbon" was also handed out to outstanding cadets. The year 1920 saw a cadet corps of the same number from the previous year. This was noted as the last year affected by the Depression of 1918 and the aftermath of the war.

In 1907, Captain Thomas A. Davis opened his first school in El Paso with a dream of creating a great military academy. A decade later, his work had been validated, the Great War irreversibly sealing the Academy's reputation as a major player on the military school stage. As Davis planned to establish his third military school in Pasadena, the Academy entered her Golden Age.

Chapter 4

The Lost Academy

The year 1917 marked a major milestone in the Academy's history: the end of the Academy's Platinum Age (1907–17) and the start of her Golden Age (1917–37). This concluded a decade from the time Thomas A. Davis opened his first school in El Paso. The year 1917 welcomed Captain Davis's younger brother as the new assistant professor of Military Science, as well as America's entrance into the Great War. This was also the year Davis would open his third school, the Pasadena Army and Navy Academy. This was the next step in his dream of a vast military school system.

The Pasadena Army and Navy Academy was probably the hardest part of this history for me to research. There was virtually no information on this Academy. No one even knew it existed. In fact, it might have been lost to history if I hadn't stumbled on a unique piece of memorabilia on an auction site: a helmet badge in the style of the San Diego Army and Navy Academy crest but bearing the words "Pasadena Army and Navy Academy—1917." Looking through all the yearbooks, articles and publications on the Academy, there was not a single note on this Pasadena Academy. Through only scraps of old newspaper articles and publications, I have been able to piece together perhaps the first history of this forgotten school.

THE PASADENA ARMY AND NAVY ACADEMY

As with the end of previous major wars, the end of the Great War renewed interest in the idea of the military school institution. Enrollment for already existing schools shot up, while officers returning from the war began establishing military schools in areas lacking one nearby. The San Diego Army and Navy Academy saw rapid growth following the war. Word of the Academy's reputation spread across Southern California. High on his recent success, Captain Thomas A. Davis decided to continue towards his dream of establishing a system of military schools based on his Davis System. The college preparatory portion of the Academy was rapidly growing; space for the already existing lower schools and proposed college was shrinking.

Meanwhile in Pasadena, there was a desire in the local community to establish its own military school within the city. Witnessing the success of SDANA nearby, the citizens of Pasadena approached Davis. They wanted to convince him to establish his next military school in Pasadena. Seeing that space at the Pacific Beach campus was running out, Davis decided that the option of establishing a new branch of his school in Pasadena would be the perfect situation for both parties. One Pasadena citizen was possibly the most excited with this new venture: Mrs. John H. Henry[9] of 1199 Oak Knoll Avenue.

Mrs. Henry was from Boston and a member of the Coolidge family, a privileged upper-class family in Massachusetts that had held influence for many generations. She grew up in the cultured atmosphere of Boston with her husband. Like Davis, she moved to Pasadena in 1910, intending to make her home there. In Pasadena, she found many other citizens who shared her exclusive perspective and saw their beautiful community as more cultured than the surrounding area. She very much desired to correct the dearth of upper-class schools near the suburbs of her residence. Having a boy of her own, she wanted the convenience of a nearby school that met her requirements rather than sending him away from home. Her goal was to purchase the land to start such a school exclusively for the education of her own son.

After the Great War, Henry and the citizens of Pasadena realized the importance of having a school that focused on the development of military character. Captain Davis was at the right place at the right time. Both Davis and the citizens of Pasadena had the same dream and goal of

9. It should be noted that Mrs. Henry took her husband's name, but he is never referred to in accompanying publications. It is assumed that he had passed away before this period. Mrs. Henry's given name is never mentioned.

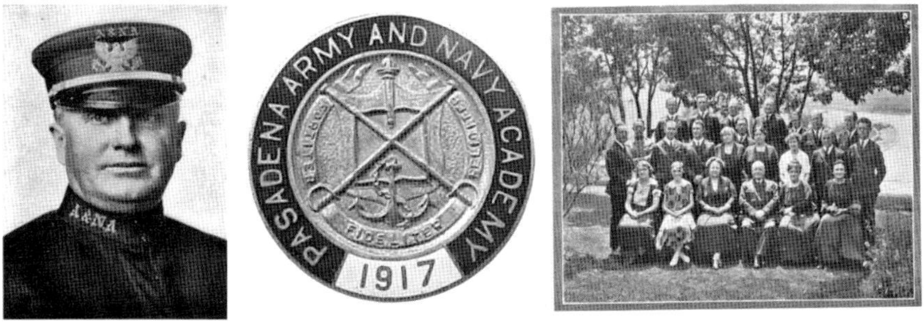

Left: Charles Wood from the 1913 *Cadet* yearbook. *Center*: Hat crest of the Pasadena Army and Navy Academy, founded in 1917. *Right*: Photograph of the Pasadena staff. *Author's collection.*

creating such a school. Davis traveled up to Pasadena to meet with Henry about establishing the Lower School branch of his school system there. After making his decision, the beautiful grounds of the former Annandale Country Club, totaling sixteen acres, were secured for the new school. This included land donated by Mrs. Henry. A new dormitory was constructed to accommodate up to forty cadets.

On October 8, 1917, the Pasadena Army and Navy Academy opened with an enrollment of eighteen cadets. As it was the Lower School branch of the San Diego Army and Navy Academy, the curriculum provided education for cadets beginning with the third grade and ending with the eighth grade. The original faculty consisted of six members, including Mrs. Henry, who owned a portion of the land; Captain Davis, who served as president in charge of Academy operations; and Colonel Charles Milton Wood, who served as the Headmaster.

Charles Wood was previously hired by Captain Davis during the tail end of the 1911–12 school year as the history and law instructor of SDANA. He left after the 1912–13 school year to join his son, who had just recently graduated from law school, to form a law practice. Davis "offered any salary" amount to Wood to remain with the Academy. Wood left nonetheless, with the yearbook noting that his departure was a heavy blow to the Academy, as he had been the "man on the job" while Davis was busy with recruiting prospective cadets. After four years, the two men were forced by fate to work with each other once more.

Marketed as a select military school for young boys of grammar school age, the Pasadena campus quickly welcomed new cadets. Henry made sure that the ads for the school emphasized the new dormitories equipped with hot and cold water, the tennis courts and the playground, as well as noting that the comfort and happiness of cadets were the first priority. At the end of the first school year (1917–18), the enrollment finished with thirty-eight cadets. The enrollment grew during the next school year (1918–19) to sixty-two. The Pasadena Army and Navy Academy became one of the top lower schools in California, and along with SDANA, was considered one of the top prep schools on the West Coast.

The publication *Los Angeles from the Mountains to the Sea*, put together by the American Historical Society in 1921 to record the growth of the Los Angeles area during this period, noted:

> *In 1910* [Captain Davis] *selected San Diego as the location for a proposed school which he founded and which was known as the San Diego Army and Navy Academy….He was persuaded by some influential residents of Pasadena to establish a branch school at Pasadena, and on October 8, 1917, opened what is known as the Pasadena Army and Navy Academy….These institutions are among the best of their kind, and America is now as never before in a position to appreciate the value of just such service as Captain Davis has rendered during his residence on the coast.*

With such acclaim, Captain Davis believed his Pasadena expansion to be a success, with the possibility of opening up a military college and other similar schools across California. However, while Davis was involved with management of the Pasadena campus, most of his time was taken up by operations at the Pacific Beach campus. In turn, Henry began to increase her influence over the Pasadena campus. Although she initially noted the importance of the military structure of the Academy, she soon wanted the school to be more than just a regular military school. Her plan for the Academy was to change the institution into an aristocratic-style boarding school meant to pamper her son and the privileged children of other wealthy families. Her vision included "perfect cuisine" for every meal, clean dormitories and a wholesome environment for the boys.

The Pacific Beach–Pasadena Schism

Although never officially stated in any of the publications, it is apparent that disagreements with Henry regarding how the school should be operated caused Captain Davis to distance himself from the Pasadena campus. Davis intended to use Pasadena specifically as the Lower School branch of SDANA, intending the cadets to transfer to the Pacific Beach campus to finish their education. Henry's ideas for the school included less emphasis on the military structure and more attention to the care of the young boys. Most important for her was a convenient local school for her son. To continue his education in Pacific Beach was too far for Henry. As her son's graduation date neared, Henry increased the concluding grade level for the school from eighth grade to tenth grade.

Around this time, most parents who had sons at the Academy strongly opposed militarism and wanted to drop the military aspects of the school. It is not hard to imagine why most of the citizens of Pasadena, who a few years earlier supported the Academy, would later oppose the manner in which she was operated. Since the Pasadena campus was just a feeder school for Davis's Pacific Beach campus, those Pasadena citizens who had invested with the Academy felt that they were losing business they could have been keeping if they dropped the Lower School and made it exclusively a nonmilitary, preparatory high school. Along with Henry, most citizens felt that the importance of a friendly environment with luxury comforts outweighed the quality of leadership and military training. Regardless of whether Captain Davis decided to leave or if he was forced out, after five semesters he left the Pasadena campus in 1919, never to return.

After Davis's departure, the school fell into the hands of Henry, who quickly changed the operations of the school to fit her ideal environment. She wanted to focus on the sweet home atmosphere that she felt was lacking at most military academies. Fancy cuisine and a clean environment were at the center of this new school. In 1920, Henry renamed the school the Pasadena Military Academy. Colonel Wood was named as the new Superintendent, filling the position formerly occupied by Captain Davis. However, it was obvious that it was actually Henry who held the reins over all campus operations.

After Henry took over, she completely rewrote the history of the school. All publications and advertisements for the Pasadena campus fail to mention Captain Thomas A. Davis as influential in the school's creation, stating instead that the school sprung "solely" from the dreams of Henry, with McGroarty's *Los Angeles from the Mountains to the Sea* noting:

> *Founded during the first year of America's participation in the World War…* [the] *Academy owes its birth and healthy growth solely to the determined*

efforts and executive ability of one woman, Mrs. John H. Henry.... There is a significance that should not be overlooked in the fact that the school from the first has been designedly military and for the education of the coming manhood of the country, and yet was founded and fostered by a woman, with no man concerned with its inception.

THE PASADENA MILITARY ACADEMY

The rewriting of the school's history after Captain Davis's departure has made it difficult to track the most accurate narrative of the new Pasadena Military Academy. Advertisements and newspaper articles after 1920 had varied histories describing the Pasadena campus. While it was known that Davis did run the Pasadena Army and Navy Academy from 1917 to 1919, this new Pasadena Military Academy tried to distance itself from Davis, resulting in conflicting histories.[10]

Now in full control of the Pasadena Military Academy, there were no obstacles left, allowing Henry to achieve her ideal vision. She spent an extensive amount of money on the improvement of buildings and equipment. Henry's expenditures reached $60,000 and included plans to build a gymnasium and indoor swimming pool, as well as a group of attractive cottages on the campus for the faculty. At the end of the 1920 school year, she dropped the elementary grades from the school and made the school strictly preparatory, with no boys under the fifth grade allowed to attend. The age range went from eight through fourteen to ten through eighteen. After dropping the Lower School, Henry stated that her ambition was for the school to be the finest of its kind in the West and that the school's reputation was yet to be made.

In an advertisement for the Pasadena Military Academy, she stated that the school was "sanely military [with] care taken, in both the Military and Athletic Departments that cadets do not exercise beyond their strength and years."

The Pasadena Military Academy offered classes in, among other subjects, business training, aeronautics and wireless telegraphy. The school was noted

10. One book that catalogues military schools mistakenly amalgamated the history of the Pasadena Military Academy and another school in Pasadena, attributed to Dr. John Ernest Blackwell, who established the Long Beach Military Academy and then later purchased in 1916 the Santa Monica Military Academy, which he operated until 1919.

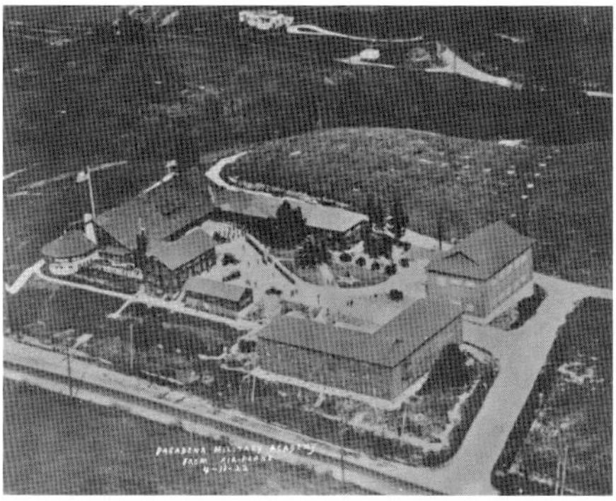

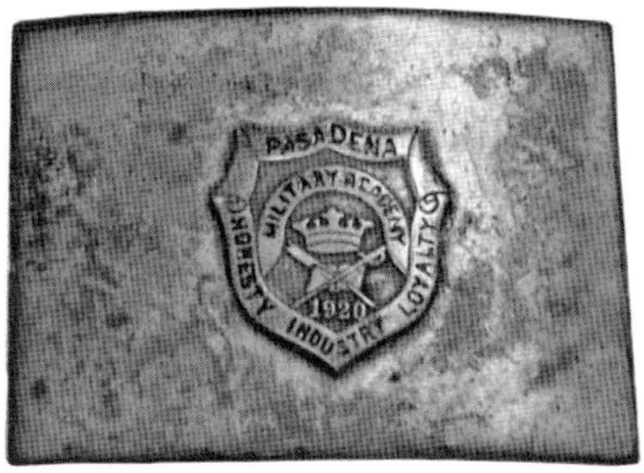

Left: Aerial view of the Pasadena Military Academy in 1922. *Right*: Pasadena Military Academy belt buckle. *Courtesy of PMA Alumni Group.*

as nondenominational, but even a casual observer would recognize the "Christian influences thrown at the cadets." Henry, who liked to refer to herself as "Godmother," stated that the school was not a reformatory school and that it was not designed for "the unselected and unclassified body of students that the public schools must serve." As noted in Sargent's *Handbook of American Private Schools* from 1922, the school only allowed "wholesome, normal, healthy boys" with a tuition of $1,200.

Finally, in 1925, the military feature of the school was dropped, and in 1926, the school moved to Covina and became incorporated as the California Preparatory School for Boys. An entirely new campus was built on the one-hundred-acre H.H. Houser Ranch in time for the opening in 1926. All that is known about this new school was that her yearbook was called *The Palm* and that the school was initially successful before falling victim to the Great Depression. Afterwards, this school limped along until 1942, when it was purchased by the Theosophical Society of Point Loma. In 1951, the California Preparatory School for Boys became the California Baptist Theological Seminary before closing its doors in 1974.

PART II

THE ACADEMY'S GOLDEN AGE

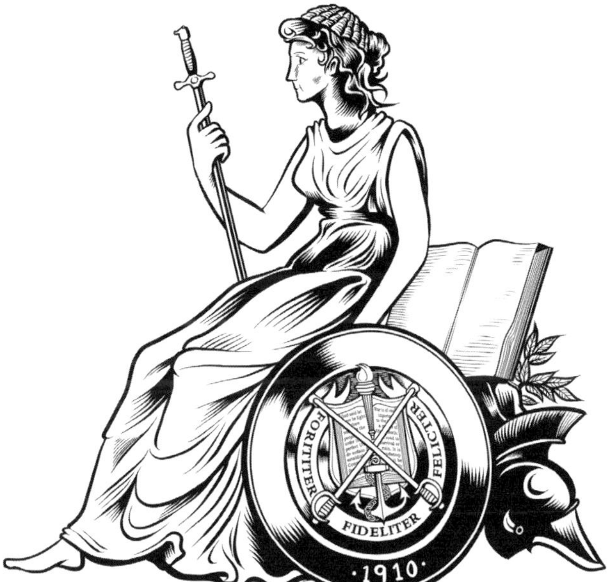

The Academy's Golden Years. Artwork by Alexander Mui.

Chapter 5
California's West Point

After the Great War, there was a boom in enrollment for the San Diego Academy Army and Navy Academy. The following period marked the beginning of what would be remembered as the Academy's Golden Age. During this era, enrollment numbers, standards and recognition were at their pinnacle. These were the years when Thomas A. Davis was proud to boast that his school was "the largest private military school west of the Mississippi River and the only school on the Pacific Coast having active membership in the Association of Military Colleges and Schools of the United States." It was during this period that the Academy was well recognized among the top pre-college schools in the nation and the forerunner of many innovations that were taking place in the military school system.

Forward March

Captain Davis was now known as Colonel Davis.[11] Following a postwar enrollment boom, the Academy quickly grew until she held claim to being the largest such school in the area. But even at this height, Davis

11. The fiftieth anniversary edition of *The Warrior*, published in 1960, notes that around this time (1920) Davis became a colonel in the Kentucky National Guard. Davis did not wear official colonel rank while partaking in Academy business and opted to wear cadet colonel ranks while serving as president. Perhaps this started the tradition of every ANA president given the honorary rank of colonel by the Board. Also, this may explain why the rank of cadet lieutenant colonel is the highest rank a senior cadet can obtain, perhaps so they would not "outrank" the president.

had many new ideas and construction plans that would expand the Academy even larger.

The 1921–22 school year was the first of eight consecutive years that marked growth for the San Diego Army and Navy Academy. That school year, the Academy was given an "excellent" rating in every department by the War Department of the United States. The Gold Academy Ribbon was created and awarded to cadets who earned the Purple Ribbon for three straight months. Another award, title "Cadet Man," was awarded to the top-performing cadet. The next year saw enrollment increased to 160. The Naval Training Program was added to the school curriculum. A bungalow near Pacific Beach Presbyterian Church was purchased for preschool classes.

A new form of the "Davis System" was introduced at the Academy. Under this evolving system, all cadets with grades of 90 or higher were listed in an honor roll, and those under 75 were placed on the "Deficiency List." Cadets on this list received personal aid from their instructors to help boost their academic standing. By the end of the year, the cadet with the greatest scholastic average was awarded a golden watch with his name inscribed, as well as the "Everett Barnes Plaque." The catalogue for SDANA during this time noted:

> *The Academy is not denominational, yet a Christian influence pervades every phase of work. Chapel exercises are held each school day and all cadets attend churches on Sunday as their parents direct. From the day of its founding clean speech, clean sports, clean life, and truthfulness and honor have been emphasized: thus boys have been taught to be thoughtful of others, to be kind and considerate, to be respectful to father and mother and gentle in old age. It gives one who loves boys deeply sympathetic with a boy's life and a peculiar happiness in securing such result.*

During this period, the Academy started annual dances known as hops, in West Point fashion. Pep rallies also became a big part of campus life, especially rallies before big games. During one very enthusiastic rally before a big game with Loyola, *The Cadet* reported, a "long line was formed with the corps in their pajamas, and they serpentine down to the Crystal Pier Ballroom, where some yells were given. The bunch then trailed back to the school, holding up traffic all the way."

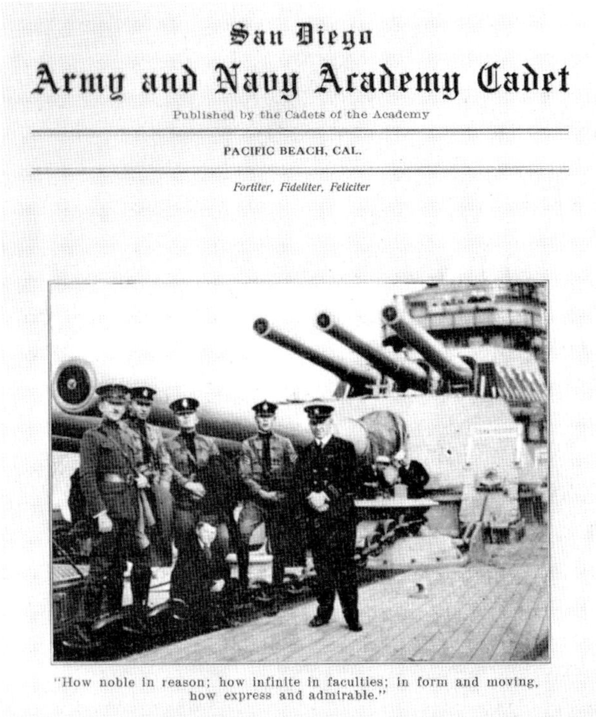

San Diego

Army and Navy Academy Cadet

Published by the Cadets of the Academy

PACIFIC BEACH, CAL.

Fortiter, Fideliter, Feliciter

"How noble in reason; how infinite in faculties; in form and moving, how express and admirable."

Right: Colonel Davis and cadets training with the U.S. Navy onboard a battleship. *Courtesy of the Army and Navy Academy.*

Below: Colored lithograph of cadets from San Diego Army and Navy Academy marching on parade. *Author's collection.*

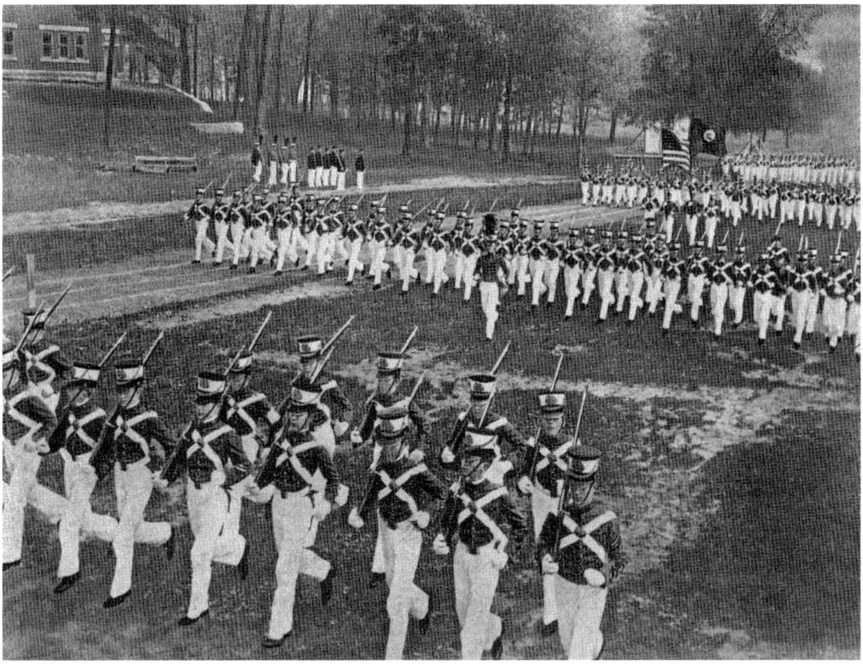

GROWING FAME

The San Diego Army and Navy Academy's football team must be mentioned when discussing the Pacific Coast Conference for the Ninth Tournament East-West Football Game in 1923. This game would come to be known as the Rose Bowl and was set to be played in the newly constructed Rose Bowl Stadium in Pasadena. An article in the November 7, 1922 *Los Angeles Times* reported:

> *Penn State's football team will positively represent the East in the New Year's Day game at Pasadena on January 1. This much was definitely settled some time ago and was confirmed in a loud tone of voice last night by Leslie J. Ayer, secretary of the Pacific Coast Conference. Hence all the efforts of the volunteer selection committees who have been choosing everybody from Washington and Jefferson to the San Diego Army and Navy Academy have gone to waste.*[12]

The establishment of the Lower School took place during the 1923–24 school year. That same year, the June 8 issue of the *Los Angeles Times* devoted a large portion of its editorial page to the outstanding work in character building that the Academy was doing. That year, SDANA joined the Association of Military Colleges and Schools of the United States, where the Academy already had the reputation of being known as the "West Point of the West."

During the summer months, Davis spent time in an office in Los Angeles recruiting cadets for the upcoming school year. This helped increase enrollment to 235 cadets for the 1924–25 school year. The construction program was renewed with more cabins, a building for the Lower School and an addition to the mess hall.

Perhaps the most important event to happen in the Academy's history, after her founding, occurred on September 24, 1925, when William C. Atkinson and his wife, Virginia "Din" Atkinson, joined the staff. Their presence made an instant impact. Virginia organized the Masque and Wig Club and presented her first play that December. Captain Atkinson became the band director, and despite having only a few old instruments to work with, that very year he trained the Warrior Band to march in the San Diego Armistice Parade. The Warrior Band won Band Unit with

---

12. *Los Angeles Times*, "Pacific Coast Eleven to Be Named on Saturday Preceding Thanksgiving Day," November 7, 1922.

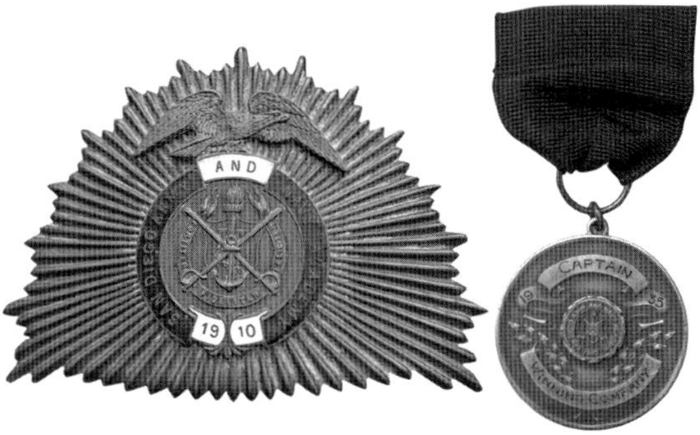

Shako hat crest from the San Diego Army and Navy Academy and the Winning Company medal from 1935. *Author's collection.*

Distinction during the parade. Atkinson also directed the band in a live broadcast over KFI, the Los Angeles Examiner Broadcasting Station, on May 22. The Warrior Band gained recognition during this time for its concerts and parades in Balboa Park.

By the 1926–27 school year, three new buildings had been completed: a new auditorium for the Drama Department, classrooms for Lower School and an infirmary. The opening of a new auditorium gave Virginia Atkinson and the Drama Department an excellent opportunity to gain recognition by putting on plays for the cadets and citizens of Pacific Beach. The enrollment rose to a cadet corps of 329. The Class of 1927 graduated with 36 graduates and 6 postgrads. A new building expansion project was also launched.

The Military Schools of California

As noted, periods after major national wars often witness a large increase in the establishment of military schools. However, after the Great War, the East and the South were already dominated by countless well-established military schools with notable reputations. To establish a new military school in the shadow of these institutions would have been pointless, as anyone who tried saw their school disappear as quickly as it was opened. In fact, only two notable military schools established during this period are still in operation: Valley Forge Military Academy (1928) and Admiral Farragut Academy (1933). The only place new schools would have had a chance would be on the West Coast, where SDANA stood as the only major competition.

The Roaring Twenties saw a boom in the establishment of Californian military schools. Wanting to mirror the success of Davis's Pacific Beach campus, numerous military schools were established around Southern California almost overnight. Most lacked a stable foundation and would last no more than a few years. Out of the fifty-four military schools in California I researched, the majority were created during the 1920s. Including the successor to Davis's Pasadena school, the Pasadena Military Institute (1920–25), notable examples included Seala Military Academy (1920–24), founded by Grenville C. Emery in Palo Alto and renamed West Coast Military Academy (1924–29); Hollywood Military Academy (1923–1940s), founded by Major Hilary A. Blackwell in Redondo Beach; South Western Military Academy (1924–25) in Long Beach, which was renamed Southern California Military Academy (1925–87); Pacific Coast Military Academy (1926–33), founded by Major Royal W. Park in Palo Alto and later renamed the Park Military Academy (1933–939) in Palo Alto; Black-Foxe Military Institute (1928–68) in Hollywood, established by Major Earle Foxe and Major Harry Lee Black on the campus of the defunct Urban Military Academy; Burbank Military Academy (1932–1940s) in Burbank, run by Lieutenant and Mrs. Demonbrun; and Pacific Military Academy (1929–46) in Culver City, established by city founder Harry Culver. The grounds of the Pacific Military Academy would serve as the campus of the later Cheviot Hills Military Academy (1946–52) under the ownership of Frank J. Brick. In 1933, Glenn Coklin, an employee at Pacific Military Academy, purchased the grounds of the Southern California Athletic and Country Club in Lake Elsinore and founded the Elsinore Naval Academy on July 29 of that year. Later, the school was renamed the Elsinore Naval and Military School and was notable for being the only other San Diego military school not under the reign of Colonel Thomas A. Davis during this era.

Some existing boarding schools and religious schools in California also decided to rebrand themselves as military schools. A notable example was Manzanita Hall, founded in 1893, which became the Palo Alto Military Academy (1919–73) during this period. None of these other Californian military schools has survived to this day, with one exception.

The only survivor was St. Catherine's Academy, which was founded by Mother Pia Backes, the founder of the Dominican Sisters of Mission San Jose, on March 19, 1889. The school was founded as an academy for girls and became a military school in the 1920s, renaming itself St. Catherine's Military Academy. The school exists as a Catholic school for young boys, kindergarten through eighth grade, and thus did not qualify for JROTC accreditation, reserved for only high schools. The school later dropped "Military" from its name and reverted to its former name in 2010.

Roughly fifty military schools have been recorded to exist in California; however, most existed for such brief periods that no information survives on these schools. It should be noted that most of these military schools were concentrated near Palo Alto and in the Los Angeles area. Grenville C. Emery was responsible for the creation of many of these schools. Emery moved in 1897 to Los Angeles, where he served as principal of the short-lived Los Angeles Military Academy in 1899. The following year, he established the Harvard School for Boys in 1900 in Los Angeles, which became nonprofit in 1911. He then later moved to Palo Alto in 1918. One of the older high schools in Palo Alto at this time was Manzanita Hall, founded by Frank Cramer in 1893 as a nonmilitary boys' day and boarding school. Emery, along with Colonel Richard P. Jelly, who worked with him at the Harvard School, turned Manzanita Hall into the Palo Alto Military Academy in 1919.

In 1920, Emery went on to establish the Seala Military Academy near Stanford University, which was later renamed the West Coast Military Academy. Major Royal W. Park, who worked at the West Coast Military Academy, went on to establish Pacific Coast Military Academy (later renamed Park Military Academy) in the same city. Both West Coast Military Academy and Park Military Academy closed before World War II, leaving the Palo Alto Military Academy the sole surviving military school in the city until it reverted to a nonmilitary school during the Vietnam War.

The Los Angeles area was home to the Los Angeles Military Academy, St. Catherine's Military Academy, California Military Academy, Harvard School for Boys, Long Beach Military Academy, Santa Monica Military Academy, Burbank Military Academy and other schools with brief existences. The more notable schools of this area were Black-Foxe Military Institute, Hollywood Military Academy and Southern California Military Academy.

Black-Foxe Military Institute was founded in Hollywood in 1928 by Hollywood developer Charles E. Toberman, along with World War I veterans Major Earle Foxe and Major Harry Lee Black. The campus occupied the former grounds of the Urban Military Academy, where Black served as commandant. Black continued as commandant of the new school, with Foxe in the position of president and Major Harry Gaver as headmaster. The location of the school attracted sons of those involved in the movie industry. Notable alumni include Charles Chaplin Jr.; Guillermo Endara, president of Panama (1989–94); Larry Marin Hagman, best known for playing J.R. Ewing in *Dallas*; Harry Carey Jr.; and Gene Wilder, who wrote that he was bullied and assaulted at the school for being the only Jewish boy.

Along with the Pasadena Military Academy, another military school was established in south Pasadena. Oneonta Military Academy was originally founded as Huntington Hall, a school for girls. At Huntington Hall, the school gardener, Harry Warren Bowen, grew jealous of his wife, a science teacher, and was discharged from the school for his manic personality. His wife described him as "depressed and melancholy....He has illusions of persecution and that people are always working against him." In a tragic turn of events, in February 1920, he shot the head of the school, Florence Housel, and then set the school on fire before attempting to kill himself. In 1922, the grounds of the former school served as the campus of Oneonta Military Academy.

The headmaster of Oneonta Military Academy's junior school was Captain Edgar H. Mohan. In 1924, Mohan moved to Pacific Beach to work as the Junior School Headmaster for SDANA. Oneonta Military Academy later moved to Culver City.

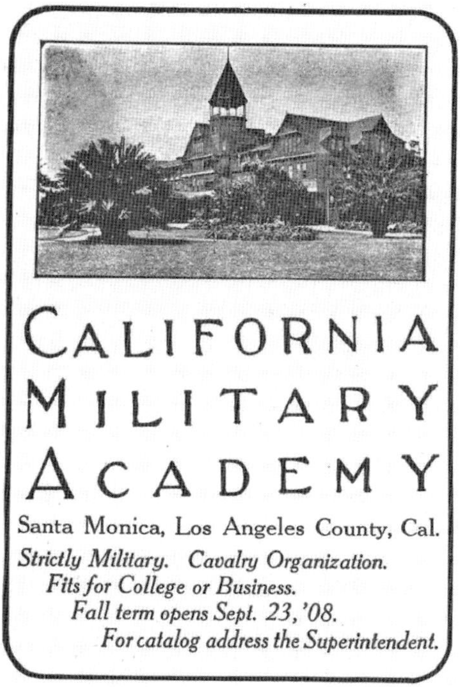

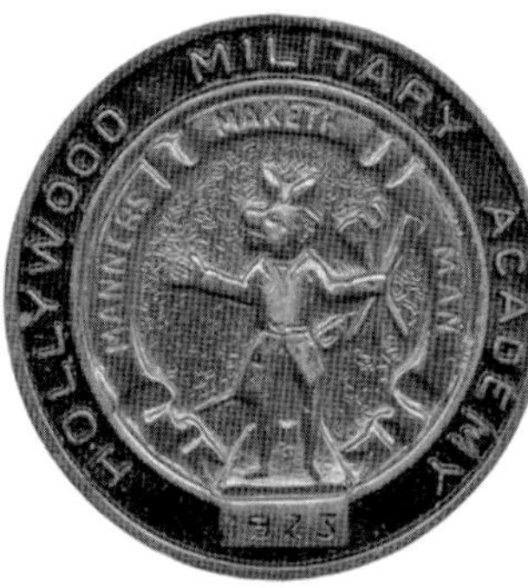

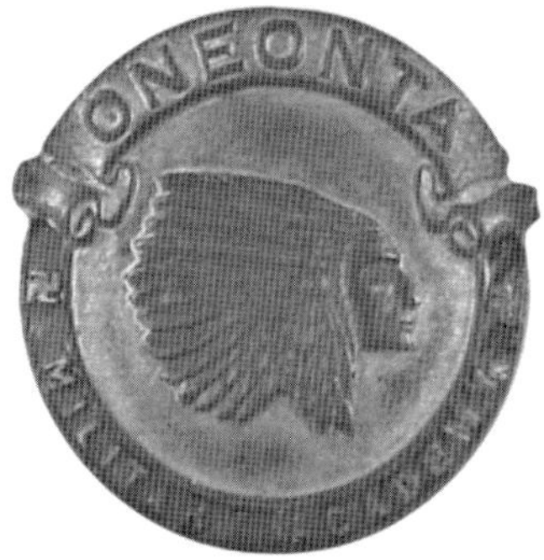

An advertisement for the California Military Academy, one of the early military schools. Badges from the Hollywood Military Academy and Oneonta Military Academy, two Los Angeles–based schools (*at right*). *Courtesy of Military Histories Blog.*

Page Military Academy in Los Angeles marketed itself as "A Big School for Little Boys." In 1941, Page was acquired by Oneonta. The two schools reorganized as Page-Oneonta Military Academy, which operated both campuses.

Major Hilary Adrian Blackwell, a British army veteran of the Great War, was the son of Dr. John Ernest Blackwell, who opened the Long Beach Military Academy and operated the Santa Monica Military Academy. Major Blackwell went on to establish the Hollywood Military Academy in Redondo Beach in 1923.

The South Western Military Academy was founded in 1924 by a group of Long Beach businessmen: C.D. Beauchamp, R.L. Stevens, I.M. Stevens, Nelson McCook and W.J. Burgin. The following year, it was purchased by Mr. and Mrs. Wilbur J. Watkins, who renamed the school Southern California Military Academy. The school was operated as a junior high military school until 1947.

Most of these schools were closed during World War II, when the government took over the property of some schools to serve as temporary military bases and facilities. Pacific Military Academy, for example, served as the military barracks for the U.S. Army's First Motion Picture Unit at the beginning of the war for "top-secret photographic work." Troops initially lived in tents on the athletic fields and would work at Culver City's Hal Roach Studios, informally called "Fort Roach." The school is immortalized in a film sequence showing a young officer, Ronald Reagan, lining up for lunch in the cafeteria. Fourteen miles south, one alumnus from Hollywood Military Academy recalled Japanese subs shelling the army barracks nearby, which resulted in the school being moved inland to Redlands. After the war, many of these government-occupied campuses were sold.

Chapter 6

Conquer and Prevail

The Junior Reserve Officers' Training Corps was created as part of the National Defense Act of 1916, intended to emulate military-style education for high school students who could not attend military schools. Under this program, civilian schools were authorized a loan of federal military equipment and military personnel as instructors under the condition that they follow a prescribed training course.

Since SDANA was the forerunner of many of the JROTC education techniques, Colonel Davis originally opted out of becoming the first California school with a JROTC program (that honor went to Grenville C. Emery's Harvard School). Davis wanted to continue developing his ever-evolving Davis System and felt that a mandatory, imposed training system would inhibit any progress. As well, the Academy welcomed students of all ages, while the JROTC program was limited to only high school–age cadets. Despite not being in the JROTC program, the U.S. Army still authorized and loaned military personnel to the Academy.

The Warrior Battalion and JROTC

In 1927, the United States War Department once again tried to convince Colonel Thomas A. Davis and Major John L. Davis to become a part of the JROTC program. They sent the Davis brothers a telegram stating that the school had earned the Class M1 military rating and was the perfect institution to help launch the JROTC program, which had expanded from just civilian schools. It should be noted that the other California military schools were trying to imitate Davis's unique education structure yet did not come close to the standards set by SDANA. Most of these military schools applied for the JROTC program and were denied.

After learning that thirty other California military schools were denied such an honor by the War Department, the Davis brothers finally allowed the Academy to become a unit within the ROTC Program. The agreement stated that Davis would still have full control over the operation of the Cadet Corps, its training and continuation of school traditions. As such, the Academy had a unique relationship with the ROTC not enjoyed by other schools.

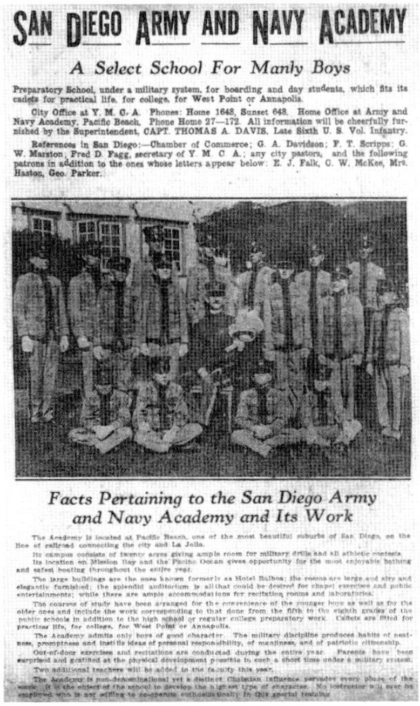

SAN DIEGO ARMY AND NAVY ACADEMY

A Select School For Manly Boys

Preparatory School, under a military system, for boarding and day students, which fits its cadets for practical life, for college, for West Point or Annapolis.

City Office at Y. M. C. A. Phones: Home 1648, Sunset 643. Home Office at Army and Navy Academy, Pacific Beach. Phone Home 27—172. All information will be cheerfully furnished by the Superintendent, CAPT. THOMAS A. DAVIS, Late Sixth U. S. Vol. Infantry.

References in San Diego:—Chamber of Commerce; G. A. Davidson; F. T. Scripps; G. W. Marston, Fred D. Fagg, secretary of Y. M. C. A.; any city pastors, and the following patrons in addition to the ones whose letters appear below: E. J. Falk, C. W. McKee, Mrs. Haston, Geo. Parker.

Facts Pertaining to the San Diego Army and Navy Academy and Its Work

The Academy is located at Pacific Beach, one of the most beautiful suburbs of San Diego, on the line of railroad connecting the city and La Jolla.

Its campus consists of twenty acres giving ample room for military drills and all athletic contests.

Its location on Mission Bay and the Pacific Ocean gives opportunity for the most enjoyable bathing and safest boating throughout the entire year.

The large buildings are the ones known formerly as Hotel Balboa; the rooms are large and airy and elegantly furnished; the splendid auditorium is all that could be desired for chapel exercises and public entertainments; while there are ample accommodations for recitation rooms and laboratories.

The courses of study have been arranged for the convenience of the younger boys as well as for the older ones and include the work corresponding to that done from the fifth to the eighth grades of the public schools in addition to the high school or regular college preparatory work. Cadets are fitted for practical life, for college, for West Point or Annapolis.

The Academy admits only boys of good character. The military discipline produces habits of neatness, promptness and instills ideas of personal responsibility, of manliness, and of patriotic citizenship.

Out-of-door exercises and recitations are conducted during the entire year. Parents have been surprised and gratified at the physical development possible in such a short time under a military system.

Two additional teachers will be added to the faculty this year.

The Academy is non-denominational yet a distinct Christian influence pervades every phase of the work. It is the object of the school to develop the highest type of character. No instructor will ever be employed who is not willing to co-operate enthusiastically in this special training.

Newspaper advertisement for the San Diego Army and Navy Academy, circa 1910s. *Author's collection.*

One of the reasons Colonel Davis finally allowed for the JROTC program on his campus was the fact that the Academy was now eligible for additional equipment, as issued by the United States government. Additionally, cadets could now become Reserve United States Army Officers upon their graduation. Major James Peyton returned to SDANA on behalf of the War Department as her first JROTC Military Instructor. Major Peyton had previously served at the Academy as the professor of Military Science and Tactics and was the inspector who gave the Academy her first government rating.

CONQUER AND PREVAIL

By 1927, the Academy consisted of sixty-four buildings and cottages for both the Lower and Upper Schools, which at this time held 362 cadets. That year, the entire Corps took a train to Los Angeles to take part in a parade to dedicate the new Los Angeles City Hall. The Academy must have made an impression, as the Cadet Corps was awarded the Silver Cup for being the best military unit in the parade and won a cup for excellence in drill, discipline, appearance and military bearing.

During the 1928–29 school year, the enrollment reached 437 registered cadets, the largest ever maintained in the history of the Academy. That fall, four new barracks were completed. The senior class had 53 cadets and 5 postgrads. With the Corps reaching almost 500 cadets, Colonel Davis initiated the largest expansion program to date.

The Academy was divided into four divisions, with Lower School representing grades first through sixth, Junior High representing seventh and eighth, High School representing ninth through twelfth and Junior College representing first- and second-year students. By January 1931, the planning of dormitories costing more than $50,000 each was completed. Construction of another three-story concrete and steel building was completed that housed sixty cadets, with an additional floor strictly for recreation and offices for the ROTC staff.

The Academy's landmark gymnasium was also built during this period. The gym served as a 150-capacity auditorium, with an indoor target shooting range in the basement. The gym was also utilized for the athletic locker room and showers. Another major landmark added

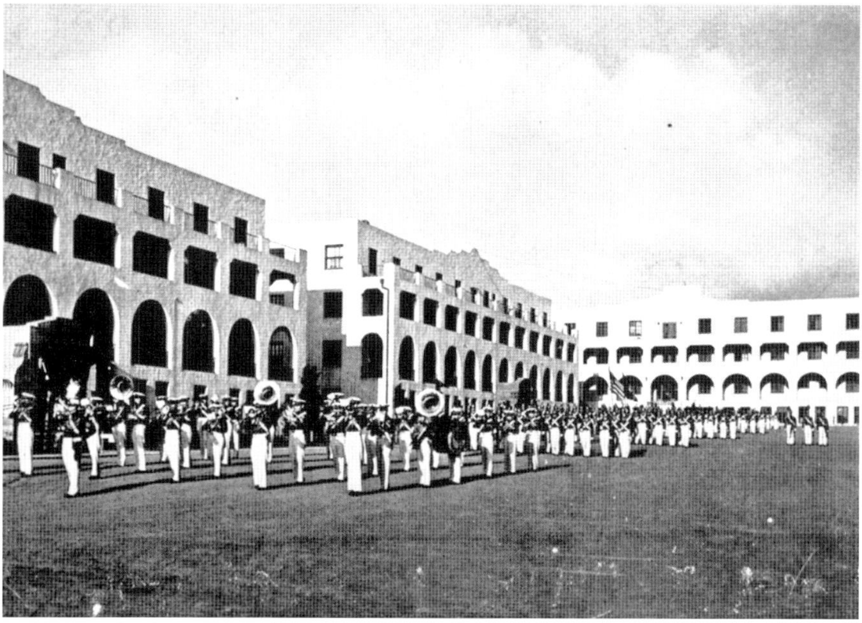

The marching band formed outside the Academy barracks. *Author's collection.*

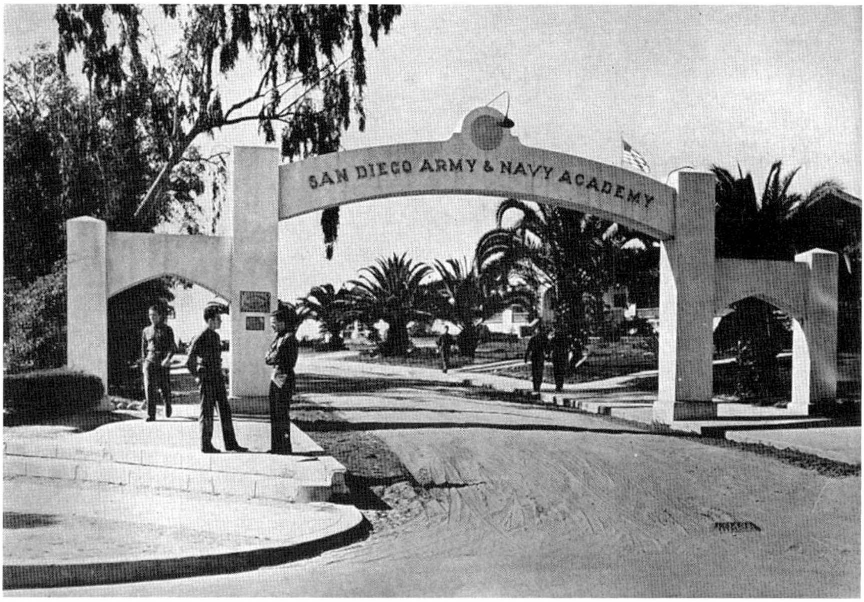

The San Diego Army and Navy Academy's Senior Archway. *Author's collection.*

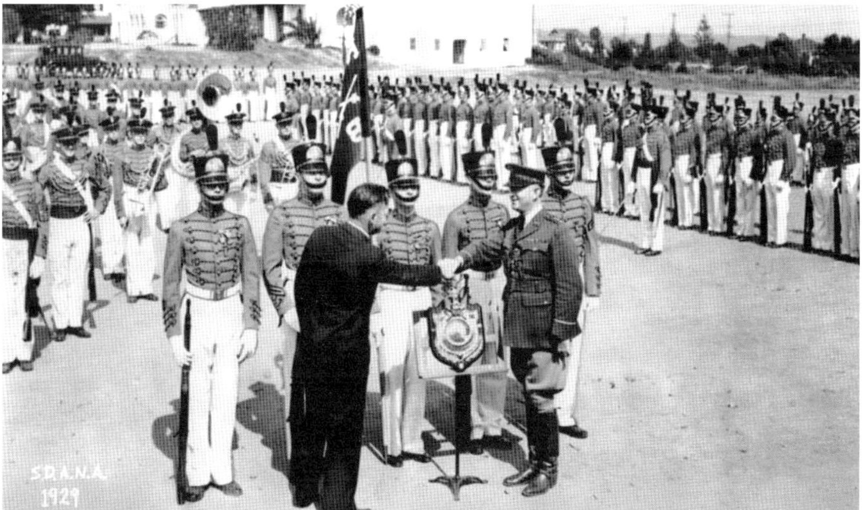

Meredith R. Pratt, on behalf of Hearst, presenting Colonel Davis and the 1929 rifle team with the William Randolph Hearst Trophy. *Courtesy of Atkinson Museum.*

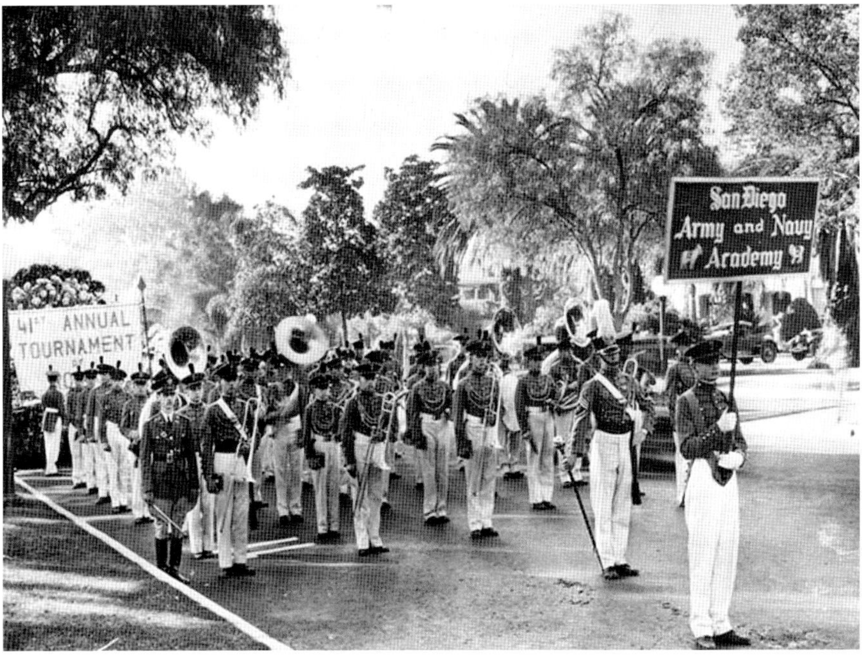

Bandmaster Atkinson (*left*) and the Academy's marching band leading the Rose Bowl Parade. *Courtesy of the Army and Navy Academy.*

to the school is the Senior Arch, which has been described in the 1930 yearbook as such:

> *Aged with tradition, a memorial to those who have served, the path 'neath this characteristic academy landmark is well trodden by those who have passed under this arch. A silent sentry of the past, an alert guard and welcome of the present, this sturdy barrier stands fast today with the promise of tomorrow. Its familiar letters in strong relief throw out a bold challenge to all who enter here, with clearly defined purpose.*

In 1929, the rifle team won the William Randolph Hearst Trophy in the Western Zone Nation Rifle Matches. The team won again the following year. Meredith R. Pratt, representative of the *Evening Herald*, presented the trophy both times to Colonel Davis on behalf of Hearst.

On New Year's Day 1930, the Cadet Corps and band led the Tournament of Roses Parade in Pasadena, a high honor, at the request of the committee in charge. Afterwards, the cadets were allowed to watch the Sixteenth Rose

Bowl Game. That same year, Band Director Atkinson wrote the Academy's first alma mater, and professor Samuel Peterson was placed in charge of the entire academic program as Headmaster.

SAN DIEGO ARMY AND NAVY JUNIOR COLLEGE

A Junior College Division might have existed within the Academy as early as 1921, although the only evidence would be from college football records. Records show that San Diego State University's first ever football game was played against the San Diego Army and Navy Academy on October 1, 1921.[13] The game was played at the Academy's Pacific Beach campus as part of the Southern California Junior College Conference (SDSU won, 6-0). The Academy was listed as a junior college for that game. San Diego State played against the Academy's High School team on October 21, 1921 (SDSU won, 12–0). The *Los Angeles Times* also mentioned the Junior College team in a 1922 article about the Ninth Rose Bowl selection. Perhaps this was simply the Academy's High School team playing at a junior college level; if so, then it is an interesting piece of history that the renowned San Diego State Aztecs got their start by competing against the Academy's High School team.

The Junior College Division of SDANA was officially established eight years later in 1929 by Captain George Fox Mott Jr., who was a major pillar of the Junior College and successfully conducted it as Headmaster for six years.[14] Mott was a graduate of Stanford University, where he got his master's degree and was professor of political science at the Academy for many years. He published the first political book on San Diego, *San Diego Politically Speaking: The Political Development of San Diego*. Later, Charles Bain served as Headmaster for the Junior College for the remainder of its existence.[15]

---

13. While it is possible the SDANA may have operated a Junior College Division as early as 1921, the only evidence is from the football records of San Diego State.
14. Different sources on the Junior College note different founding dates. The main claim is that George F. Mott founded the Junior College in 1929 as a separate branch of the Academy. It should be noted that this information comes from a letter that accompanied Junior College memorabilia donated by Mott to Atkinson decades later. The 1930 yearbook supports this claim, as it notes that the 1929–30 school year was the first year the Junior College was in operation.
15. The 1936 yearbook notes that Charles Bain operated the Junior College as Headmaster for twelve years by this point, which would have meant that the Junior College was founded in 1924, contradicting the history reported in the early 1930 yearbook.

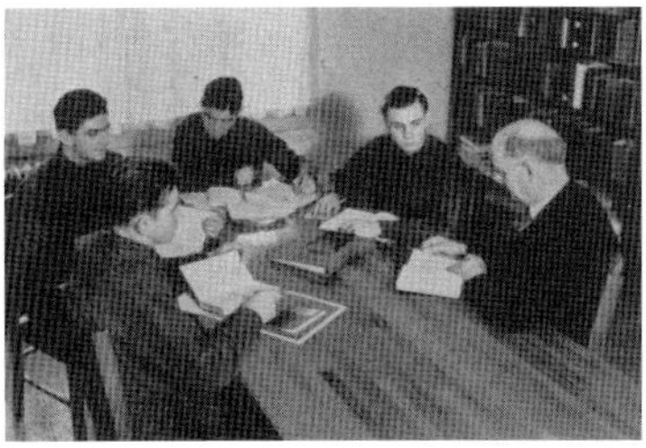

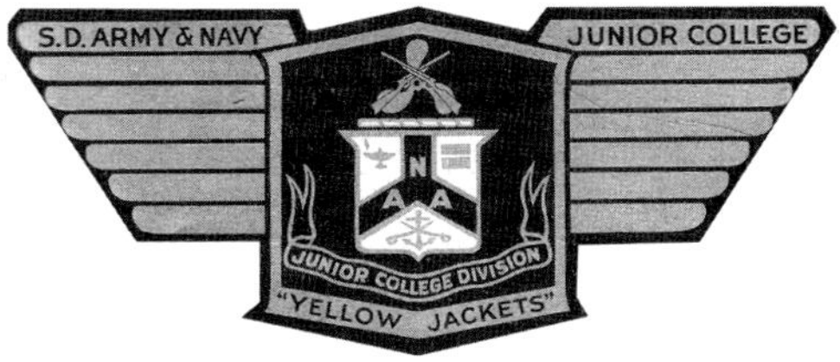

Peterson, Headmaster of the Junior College, teaching cadets of the Junior College in 1935, along with a decal sticker for the Junior College with the nickname "Yellow Jackets." *Author's collection.*

Colonel Davis purchased an additional eighteen acres to accommodate the college, which would include the construction of an entirely separate campus. When the Junior College first opened, it was the fastest-growing division within the school system. The college was unique in that it combined the operations of a junior college with that of a military university. The San Diego Army and Navy Academy Junior College was unique in that it operated independently from the rest of SDANA.

In the 1930 yearbook, Corporal Murray noted, "The first college course in the history of the Academy was offered this year, and from the very outset it certainly justified its foundation. Due to the small size of the class, which numbered only fifteen members, it was invited to affiliate with the Seniors, but it decided to stand alone."

The first class was made up of sixteen young men—eight graduated from the High School Division, while eight came from other high schools. Major Peterson served as the faculty advisor to this class, and cadet Captain del Valle was voted the class president. A banquet and dedication ceremony was held as Colonel Davis and Major Davis welcomed the first class of this new branch of the school. The members of this first class were referred to as the "Exclusive 30." When Stanford University issued two standard tests, the Academy's freshmen class ranked first out of forty-three other schools in the state. Because of its high academic standing, the Junior College Division was able to transfer many of its students to leading colleges of the nation. Along with the three major service academies, many of the students attended Stanford University, University of California–Berkeley, University of California–Los Angeles, University of Southern California, Pomona College, University of Arizona, University of Washington, Yale, General Motors School of Technology and Miami University. By 1935, the Junior College was only one of twenty private junior colleges in the entire state.

The Junior College sports team was called the Yellow Jackets. The Yellow Jackets would go on to play football against other colleges on the West Coast, including Stanford, Fresno, San Jose State and Oregon State. However, due to its small size, the Junior College football team initially ranked at the bottom of the list; within a few years, though, it would become a formidable junior college football team, with players said to be some of the greatest "seen in some time" and "without equal in junior college ranks." In 1928, the Junior College team gave the Los Angeles Loyola Cubs, the oldest high school in Southern California and which had previously been undefeated for two seasons, their first loss in two years with a 25–0 victory. San Diego State University would face the Academy one final time when it defeated the Junior College team on October 6, 1934, with a 7–19 win.

One notable alumnus from the Junior College was Lieutenant General George I. Forsythe. Forsythe was born in Butte, Montana; he graduated from Butte High School in 1935 and then traveled all the way to California to attend the San Diego Army and Navy Junior College. He graduated from the Junior College in 1937. His time at the Junior College awarded him the rank of second lieutenant in the army. He served active duty during World War II and was promoted to brigadier general in 1963, eventually reaching the rank of lieutenant general in 1966. Forsythe was selected by Secretary of the Army and Army Chief of Staff General William Westmoreland to serve as "special assistant" when there was a need for fundamental changes within the army during the Vietnam War.

General Westmoreland placed Forsythe in charge of the Modern Volunteer Force, where he helped rebuild the army by committing to the "zero draft" policy, stopping reliance on draftees. Forsythe was described as a "renegade brigadier who could do a hundred things at the same time with complete indifference

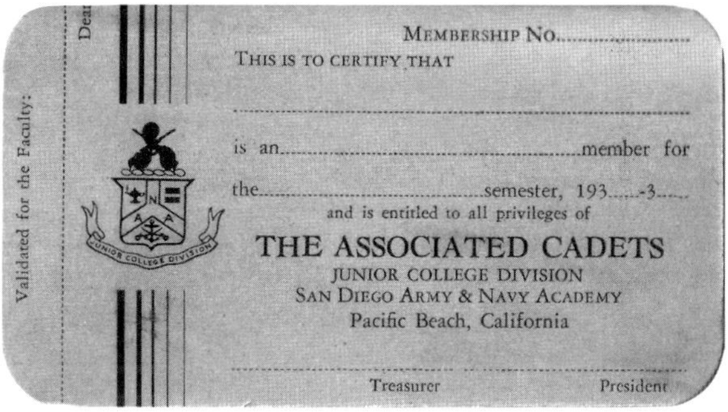

Validated for the Faculty:
Dean

MEMBERSHIP No..........................

THIS IS TO CERTIFY THAT

....

is an..member for

the..................................semester, 193.......-3.......

and is entitled to all privileges of

THE ASSOCIATED CADETS

JUNIOR COLLEGE DIVISION

SAN DIEGO ARMY & NAVY ACADEMY

Pacific Beach, California

Treasurer President

Associated Cadets membership card for the Junior College Division. *Courtesy of Atkinson Museum.*

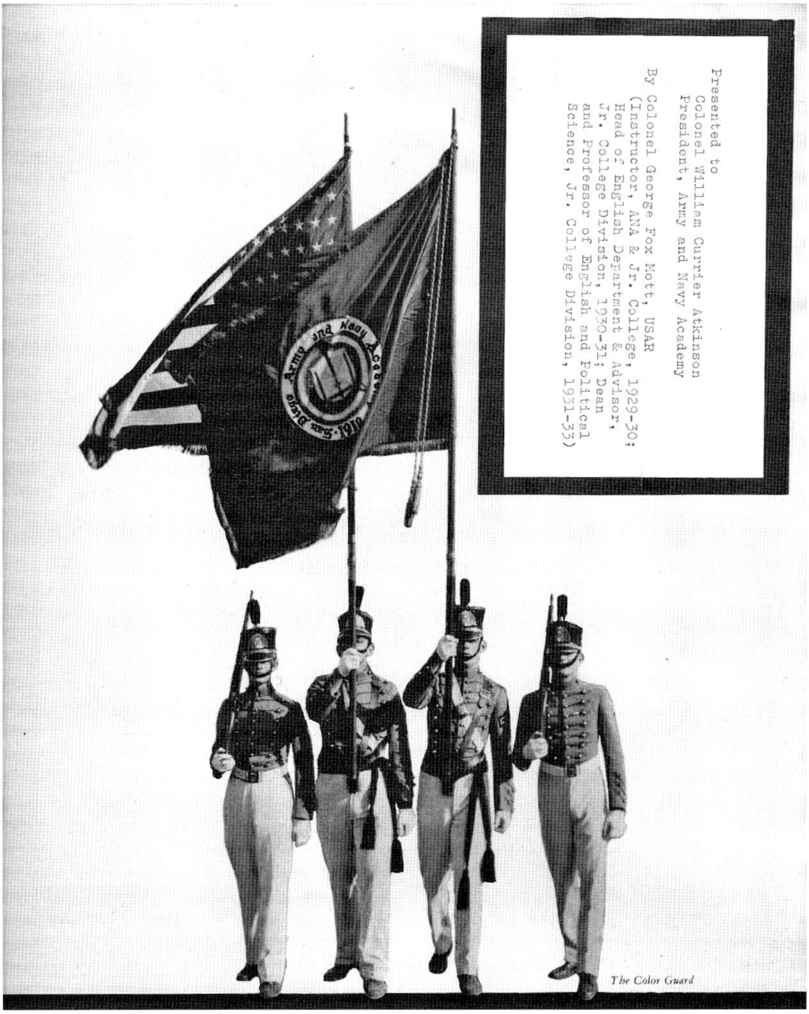

Page from the 1935 San Diego Army and Navy Academy catalogue donated by Colonel George Fox Mott, head of the Junior College, to Colonel Atkinson. *Courtesy of Atkinson Museum.*

to politeness and customary military propriety"—hence, he was perfect for handing the army's bureaucracy and Congressional forces while restoring the professional standards of the U.S. Army and its ability to fight and win wars. The adaptation of an all-volunteer force was seen as a major step forwards for the army during this period. General Westmoreland would later describe Forsythe as the "Father of the Modern Army."

LARGEST WEST OF THE MISSISSIPPI

The year 1930 marked the San Diego Army and Navy Academy's twentieth anniversary. The graduating senior class that school year, Class of 1931, numbered sixty-two cadets. In the catalogues and yearbooks, the Academy proudly boasted her motto, "West Point of the West," and publicized herself as the fifth-largest school of its kind in the nation and the largest private school west of the Mississippi.[16] A 1931 catalogue noted, "From its very small beginning the Academy grew until today it is the largest private military school west of the Mississippi River, and is the only school on the Pacific Coast having an active membership in the Association of Military Colleges and Schools of the United States."

In 1931, Colonel Davis released plans to the press that he was going to initiate another building program that would double the existing student capacity from five hundred to one thousand. New course activities for cadets were introduced during this period, including aviation courses and flying lessons, different music instrumental lessons and even horseback riding. Sports were an important part of cadet life. The Academy played against both military and civilian schools and colleges in the area. Athletically, the High School was rated among the highest in the San Diego leagues at that time.

Despite the increased student body, Colonel Davis also made sure that each cadet got enough personalized training and care. In military, physical and academic aspects of life, Davis made sure cadets were motivated to reach their full potential. Discipline was very important in the Academy, and any cadet not able to reach Davis's standards was expelled. Hazing was prohibited. Each cadet was to be responsible for himself, his teammates and everyone in his company. The Academy remained strict on her military structure, forcing cadets to wear West Point uniforms at all times. Civilian clothes were strictly prohibited from campus.

During this period, a typical day at the Academy began with reveille at 0620, followed by army exercise, inspection of uniforms and then first mess. Classes took place from 0800 to 1200, followed by lunch and military drill. From 1500 to 1800, sports practice and PT went on until dinner. This was followed by study hall and taps at 2115.

The Academy was at the pinnacle of success in its Golden Age. Nevertheless, all good things eventually come to an end. By the end of

16. These types of claims are hard to verify, as there is no official record listing the size of the different military schools across the nation. However, this is a claim quoted on numerous occasions in catalogues and advertisements throughout the early 1930s concerning SDANA.

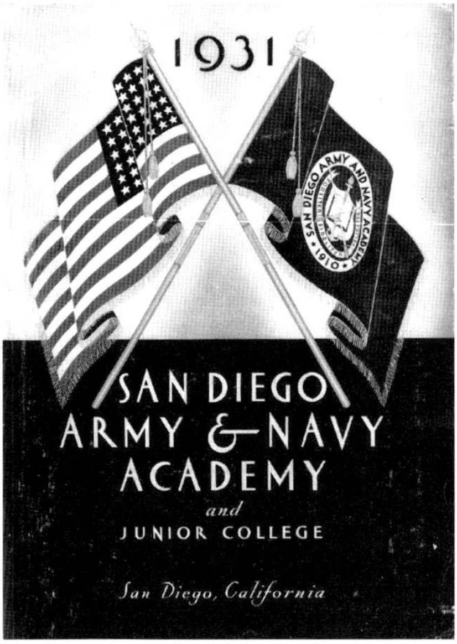

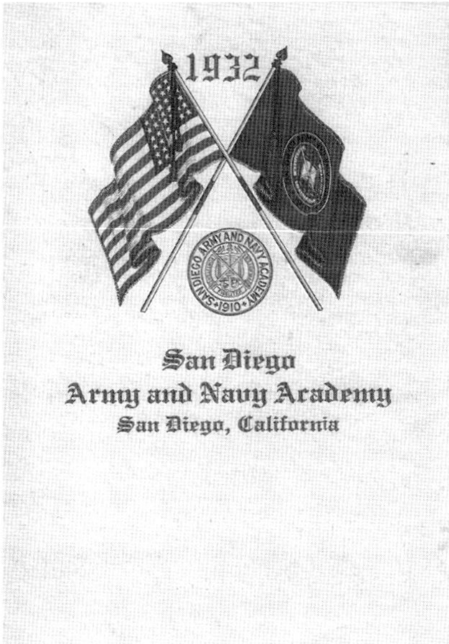

Left: Cover of the 1931 San Diego Army and Navy Academy and Junior College catalogue. *Author's collection.*

Right: Detail of the Academy's 1932 yearbook cover. *Author's collection.*

the 1931 school year, the rumored economic problems of the East were felt by the Academy. The Great Depression finally hit California. Within a year, the Cadet Corps dropped by nearly 150, and enrollment for that year numbered 253.

Even with the troubling Depression, two new clubs were formed: the National Honor Society and the Cosmopolitan Club, for the growing number of foreign students. Under Virginia Atkinson, the famed Masque and Wig Club continued to win first place in the one-act play competition at the Pasadena Playhouse year after year. During the 1932–33 school year, two cadets from the Philomatheon Society won first and second place in the National Oratorical Contest.

The Depression lingered, but the Academy pressed on, with the Corps dropping to 185 cadets. Colonel Davis and Major Davis, the sole owners of the property, were forced to reduce all salaries in an effort to meet obligations, as well as making every effort to keep expenses to a minimum. On June 23 of that school year, 67 senior cadets graduated. By 1932–33, enrollment had dropped to only 133. In 1933, the Security Trust and Savings Bank and other creditors assumed full management of SDANA under a Board

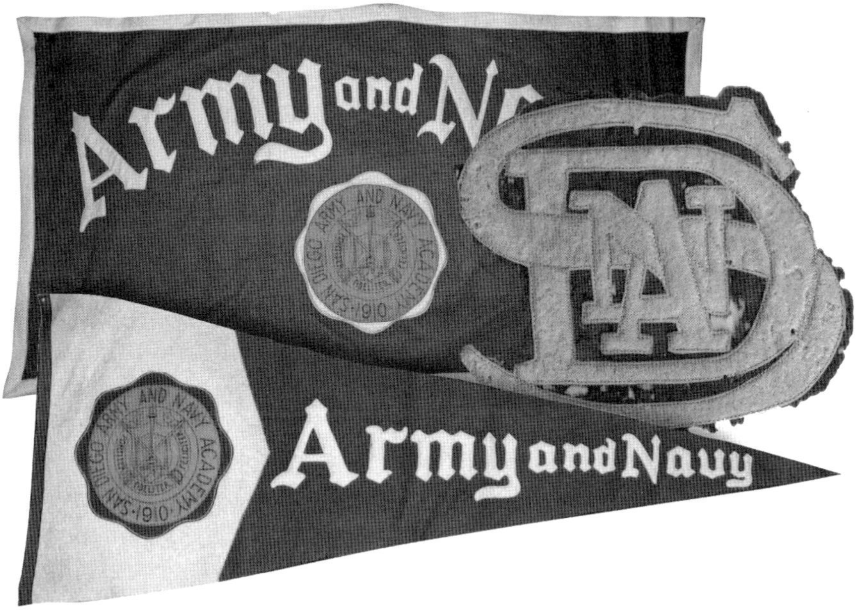

San Diego Army and Navy Academy sports letter; the colors of the letter are the Academy's original school colors, purple and white. *Courtesy of Atkinson Museum.*

of Trustees, with both Colonel Davis and Major Davis as members of the Board. By 1933–34, Major Peyton had been forced to retire for health reasons, and enrollment steadied at 132 cadets. This was a drop of more than 300 cadets within the last three years. Despite financial difficulties, Colonel Davis ensured that the Academy would continue on with the same high standards she had for decades. Then, in 1934–35, the Corps rose to 174 cadets, a sign of hope.

Chapter 7
The Black Year

The year was 1935. The United States was recovering from the Great Depression. President Roosevelt just signed the New Deal. And *Dinky*, a movie starring Jackie Cooper and filmed on the San Diego Army and Navy Academy campus, was released in theaters.

In *Dinky*, Cooper played the title character, a cadet at military boarding school. Actress Mary Astor played his mother, a secretary who was framed in a bond scam and sent to jail. Cooper was a twelve-year-old child actor known for trying to act like the tough kid. During a football fight scene filmed at the Pacific Beach campus with actual enrolled cadets, Cooper insisted the fisticuffs be real. "I don't want fellows like these to treat me like a sissy!" Cooper said of the cadets. Astor wrote that when the cadets refused to throw real punches, Cooper appeared visibly angry and ran crying to his mother, saying, "I *hate* being a movie actor—I *hate* it! I want to be a guy, a man!"[17]

End of the Golden Age

In the fall of 1935, the San Diego Army and Navy Academy celebrated its twenty-fifth anniversary, with the Corps having grown to 225 cadets, a hopeful sign that the effects of the Great Depression had come to an end.

17. From the Turner Classic Movies entry on *Dinky* (1935), directed by D. Ross Lederman and Howard Bretherton.

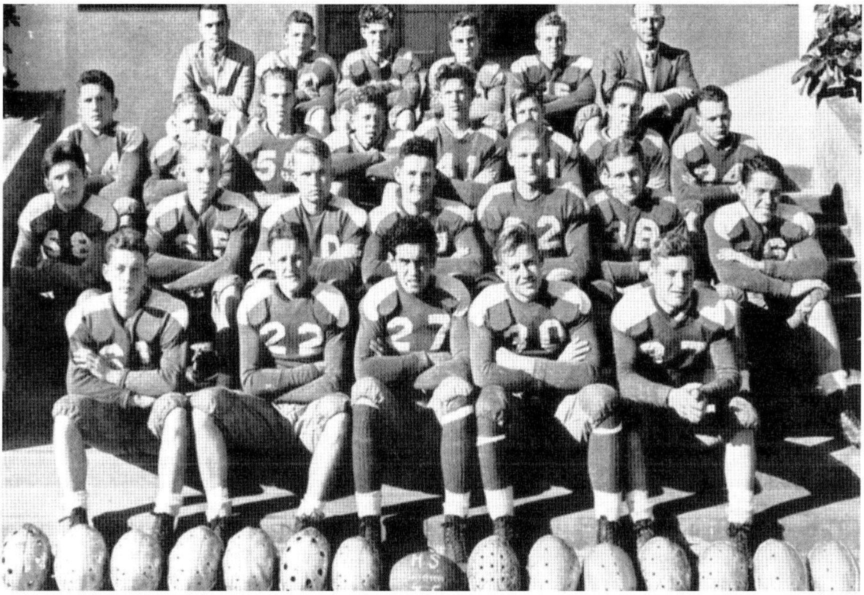

The 1935 Junior College football team, including members of the All Southern Players. Team captain Miller is seated in the front center. *Author's collection.*

That year's High School football team was considered the "dark horse" of the Metropolitan League. The rifle team brought home the William Randolph Hearst Trophy for the first time since 1930. And the Junior College basketball team had an impressive comeback year. Its most impressive victory was against the Marine Corps team, a team that had remained undefeated for three years prior and even placed in the Olympic tryout.

That year, five members from the Junior College football team were selected on the legendary "All Southwestern Players" team. These included Captain Kess Miller, Larry Meikle, Harold Throneson, Don MacDonald and Jim Barney, who were popular players from the 1935 season. Miller was chosen almost unanimously by the board of Junior College coaches and heralded as the "greatest passer seen in some time." The 1936 edition of *The Cadet* noted:

> *In choosing Miller…this selection board made no mistake…in six games he had 73 of his accurately thrown passes complete. His blocking and running ability were without an equal in the junior college ranks. In captaining this year's team, Miller has proven his line field generalship and has been an inspiration to his team.*

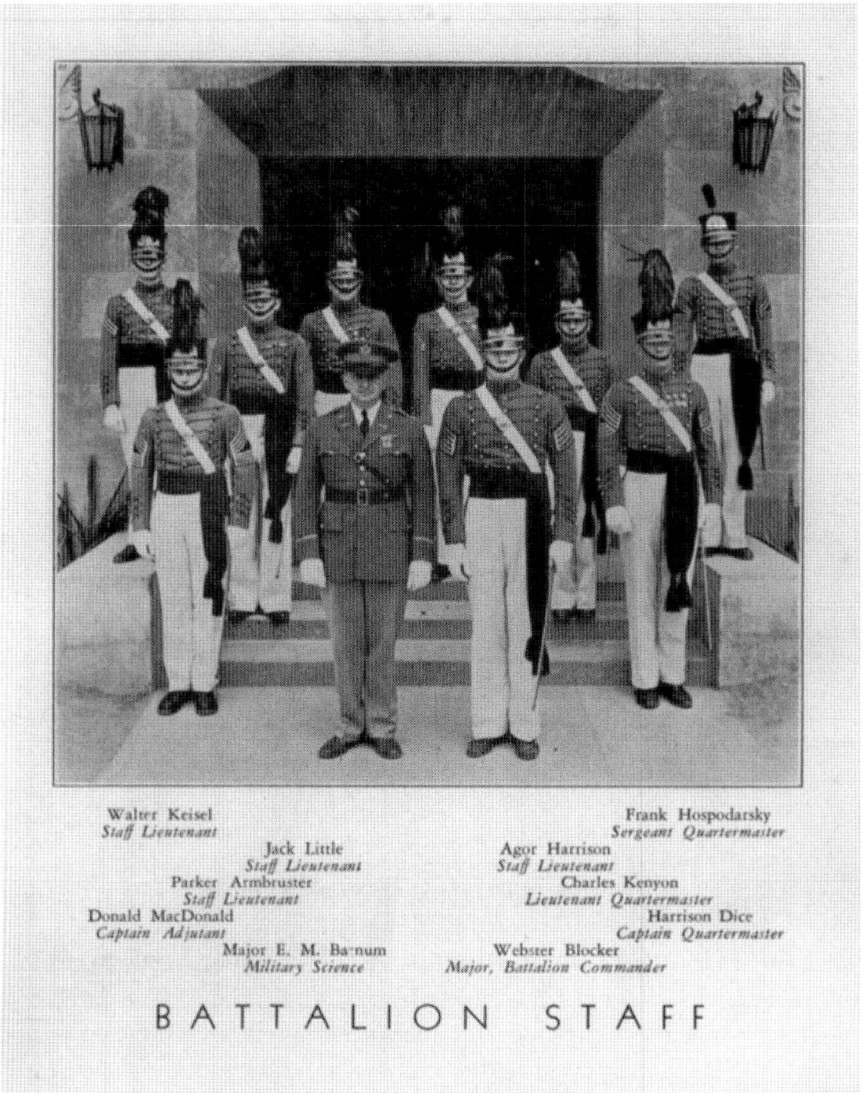

The Battalion Staff during the 1935–36 school year. Major Barnum, the instructor of Military Science, is standing in the front center with cadet Major Webster Blocker, the last battalion commander of the San Diego Army and Navy Academy's Pacific Beach campus. *Author's collection.*

That New Year's Day, under Captain Atkinson, the Warrior Band and Academy orchestra once again marched in the Pasadena Rose Parade. Things were looking up for SDANA. The Academy had basked in the success of her Golden Age, strides ahead of her peers; however, all eras must eventually give way to the next.

This period was referred to as the "Black Years."[18] Everyone feared the worst, although many thought that the impossible would never happen, as the Academy was regarded as one of the finest preparatory schools in the country and the most prosperous of the California military schools. However, the 1935–36 school year took an unexpected turn. Colonel Davis and Major Davis, who were minority members on the Board of Trustees, received an official notice from the other board members that stunned them. The Security Trust and Savings Bank and other creditors would take complete ownership of the entire property and all the Academy's assets. They did allow Colonel Davis to run the school for the time being while they figured out how they wanted to proceed.

Colonel Davis tried to keep the takeover by the other board members secret from the press and public. He continued running the Academy like normal and took care to keep everyone's minds off the Depression. Davis masked the issue of the Academy's finances with talks of the twenty-fifth anniversary celebrations. As part of the celebrations, he announced new building plans and expansions. Staff, parents and cadets were too busy preparing to pay any mind to the real programs concealed to them. To them, if SDANA had survived for a quarter of a century, then she would survive for another quarter more.

TAPS FOR PACIFIC BEACH

In May 1936, Colonel Davis announced that plans were underway for refinancing and enlarging the school with the help of federal aid. Nonetheless, within three months of this announcement, the Davis family left Pacific Beach. The press had a field day with this event and speculated that the colonel had been asked by the Security Trust and Savings Bank to surrender his control of the Academy. However, when asked, the bank released a simple press release stating, "Influences of the Depression led to lower enrollment and there were not enough funds to pay expenses associated with the building projects." Publically, Davis was forced into foreclosure by the banks, which officially assumed all aspects of campus operation.

18. "The Black Years" is a phrase used in the fiftieth-anniversary edition of *The Warrior* to describe the period after the foreclosure of the Pacific Beach campus and up to the departure of Colonel Davis.

In *The Cadet*, Colonel Thomas A. Davis expressed his feelings as he gave the Class of 1936 his final address:

> *Into a world still disturbed with a thousand problems, social, political, economic, and religious, into a world still as difficult and interesting I send you forth with silent prayer. The Future is yours. May you have the grace, courage and wisdom to meet these problems, and to help toward the realization of what God meant this beautiful world to be.*

The San Diego Army and Navy Academy—the largest military school in California, the fifth-largest private school in the nation and the largest school of its kind west of the Mississippi—finally shut her doors. The sun was slowly setting over the campus as the cadets of SDANA marched off the field for the last time.

Harry Stuart, Class of 1936, was the last surviving member of the last class to graduate from SDANA. He went on to become a pilot with American

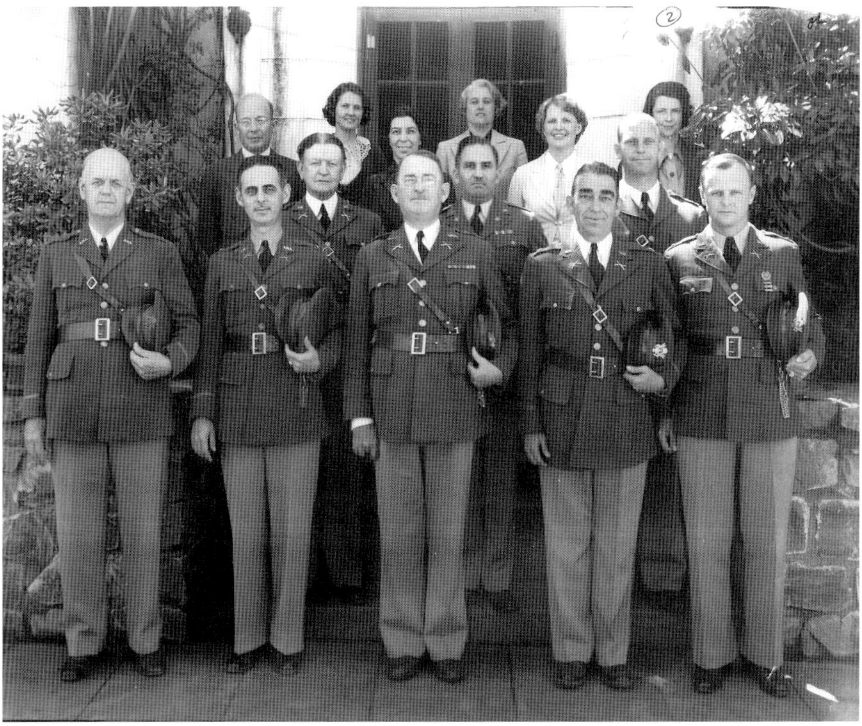

Staff of Pacific Beach who followed Davis to Carlsbad. *Courtesy of Army and Navy Academy.*

Airlines until his retirement in 1978. Recalling his time at Pacific Beach, he said, "I liked the military routine and discipline, it was good for me, and I made many friends."

Colonel Davis kept his proud reputation and said to all the alumni, cadets and parents that there would be a new start, with a new school. Davis and the other staff members had no money to purchase new land, so he presented an offer of stock shares to the department heads and other friends of the school who would join him on a new venture. The Academy then became owned by a private stock company consisting of his brother, Academy department heads and himself.

Colonel Davis learned of two pieces of property that had become available: the Samarkand Hotel in Santa Barbara and the Red Apple Inn in Carlsbad. The Santa Barbara site was a long two-hundred-mile journey from the Pacific Beach campus; however; the Carlsbad site was just thirty-five miles north. Seeing that the Red Apple Inn property included oceanfront property, he selected this site as the location for his new campus. Fortunately, the same factors that forced the Academy to close allowed this land to wind up on the real estate market. Davis signed a lease with the owner of the inn. With eleven of the faculty members from the San Diego Army and Navy Academy, he opened the Davis Military Academy with thirty-seven cadets. The slogan for Davis's Academy was changed from "West Point of the West" to "For Thee Shall Be Saved."

PART III

SHELTERED BY THE SUN-KISSED HILLS

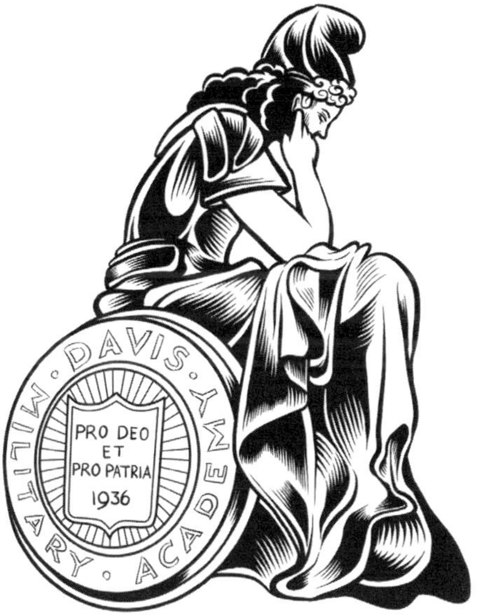

The Academy's Interregnum. Artwork by Alexander Mui.

CHAPTER 8

BORDERED BY THE SEA

FORMER SEA CAPTAIN JOHN FRAZIER was digging a well at his farm in northern San Diego in 1882. When his digging reached 415 feet, he found something that puzzled him. Frazier started digging another hole, this time 510 feet deep. He had discovered an aquifer of mineral water.

Frazier believed the water to be very therapeutic and hold health benefits. He began selling the water at the local train station, later named Frazier's Station. When analysis of the water showed that it was similar to the mineral water from the world-famous Bohemian Karslbad heath resort, today known as Karlovy Vary in the Czech Republic, he began marketing the water as Carlsbad Mineral Water. With a group of other businessmen—including Gerhard Schutte, Samuel Church Smith, D.D. Wadsworth and Henry Nelson—the Carlsbad Land and Mineral Water Company was formed.

A small community grew around the original well, which is still preserved today at Alt Karlsbad Mineral Water Spa on Carlsbad Boulevard. In the 1880s, hotels, spas and homes popped up near the well. The town quickly became a popular tourist destination. Movie stars and prominent figures were attracted to the famous mineral baths and spas. Hotels like California Carlsbad Mineral Water Hotel, Twin Inns, Red Apple Inn and Los Diego Hotel were ideal midway rest stops for tourists traveling between Los Angeles and San Diego. The city would be known as Carlsbad, in honor of its mineral water.

THE RED APPLE INN

In 1923, near Frazier's first well, A.C. Blair purchased forty acres of land along Carlsbad Boulevard. In 1927, he designed and built a twenty-room hotel called the Red Apple Inn. The inn opened for business on July 1 of that year.

Apples were the theme. They were on all light fixtures, furniture, china, doorknobs and uniforms. The red tiled roof was an eye catcher to tourists traveling down the historic Highway 101; many would stop in for a bite. The dining room served apple-themed meals served by girls in apple-themed uniforms and became famous for, of course, the apple pie. The restaurant, which thrived during its heyday, was managed by Margaret Drought, who had already run several successful restaurants. As with her previous restaurants, Drought did not employ men at the Red Apple Inn.

Nine years later, the Red Apple Inn fell prey to the Depression and closed its doors. Although only lasting less than a decade, the inn was best remembered by those who passed through for its delicious apple pie. A plaque was later placed on the building by the Carlsbad Historical Society commemorating the significance of this small hotel.

THE DAVIS MILITARY ACADEMY

During the Depression, the small Carlsbad town became a financial dump. The famous hotels and mineral spa were forced to close, and real estate prices took a sharp decline. When Colonel Davis and his family arrived, the citizens welcomed them as a ray of hope. Carlsbad did everything it could to welcome the Davis family. As the town saw it, a military school run by the popular Colonel Davis would help raise real estate prices. In August 1936, Blair gladly rented his landholdings in Granville Park to Davis, who signed a ten-year lease.

Colonel Davis arrived with, it is believed, eleven members of his staff from SDANA who wanted to follow him in his new venture. Along with the other stockholders, the Davis family and Academy staff spent the entire summer cutting weeds, cleaning the buildings and preparing the way for a new campus.

On September 16, 1936, Colonel Davis greeted thirty-seven cadets on the opening day of the Davis Military Academy. While Davis could not use the original name due to the bank owning the rights, he saw this new campus as a continuation of SDANA. In 1937, when evangelist John Brown bought the Pacific Beach property, Davis immediately registered the school under its original name, and the Davis Military Academy once again became San Diego Army and Navy Academy.

On opening day, the local paper reported that fifty cadets in uniform were present for the dedication ceremonies, with bandmaster Atkinson conducting the cadet band. The national anthem was played as the color guard raised the flag over the new campus. Colonel Davis gave a speech thanking the local community. Attending guests included C.D. McClellan, representing the Carlsbad Chamber of Commerce; Ted Holden, representing the Oceanside Chamber of Commerce; representatives of the Carlsbad Rotary Club and Oceanside Kiwanis Club; the state highway commission; and members of the schools of Carlsbad and Oceanside. Drama coach Virginia Atkinson was given an unusually big round of applause as she spoke of plans for the community theater.

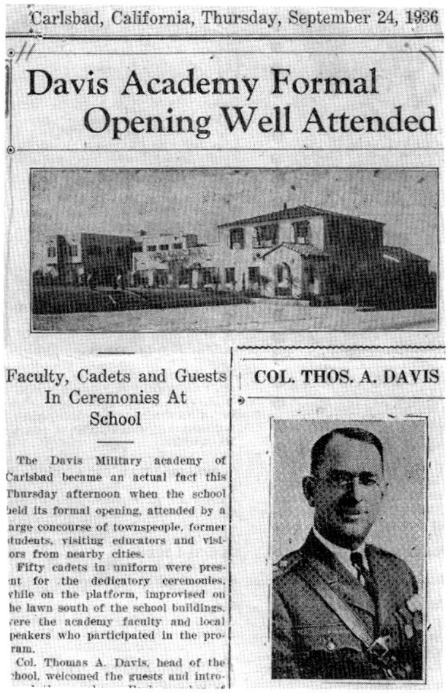

Carlsbad, California, Thursday, September 24, 1936

Davis Academy Formal Opening Well Attended

Faculty, Cadets and Guests In Ceremonies At School

The Davis Military academy of Carlsbad became an actual fact this Thursday afternoon when the school held its formal opening, attended by a large concourse of townspeople, former students, visiting educators and visitors from nearby cities.

Fifty cadets in uniform were present for the dedicatory ceremonies, while on the platform, improvised on the lawn south of the school buildings, were the academy faculty and local speakers who participated in the program.

Col. Thomas A. Davis, head of the school, welcomed the guests and intro-

COL. THOS. A. DAVIS

Article covering the Davis Military Academy's opening day. *Courtesy of Atkinson Museum.*

Many cadets followed Davis to the new campus. Some continued to wear their old SDANA varsity jackets. The Class of 1937 was the first graduating class of the new campus; they claimed the Red Apple Inn's rose garden and little stone well as "senior property." Only the upperclassmen were allowed to step foot on senior property; any lowerclassmen caught in this area were dunked in the well. The Senior Garden was the first maintained upperclassmen tradition of this new campus. One tradition that continues to this day states that a bucket must always be hanging from the well, never to be removed or taken down. Another tradition states that the gate to the garden must remain locked all year, only to be opened during graduation day.

The tuition for the new school was $1,000 per boarding student. To make room for the cadets, Colonel Davis leased the Los Diego Hotel located on State Street and Grand Avenue to serve as housing and classrooms, with the lobby to be used as the band room. The Los Diego was built in 1925 and was a famous rest stop between Los Angeles and San Diego. Its name is a combination of Los Angeles and San Diego in honor of the two cities. The Carlsbad Theatre was used by the Drama Department. The theater was constructed in 1926 at a cost of $40,000 and could seat six hundred people.

DAVIS AND DORMAN

A.C. Blair needed money to free himself from debt that resulted from the Depression. In 1938, he sold the Red Apple Inn property and mortgage to Arthur C. Anderson and his wife, Gertrude H. Anderson. The Andersons introduced themselves to Colonel Davis, expressing their interest in the work he was doing. The Andersons were known for helping youth organizations, had no children, were old in age and had a large amount of money; they became the Academy's biggest supporters. They allowed the campus to expand into their land adjacent to the Red Apple Inn, including an area known as Granville Park originally planned as a neighborhood subdivision in 1922. They then sponsored the construction of new dorms on their property. The Andersons had an admiration for cottages, so seven four-room cottages were constructed that year. Each cottage held eight cadets.

The Andersons then introduced Colonel Davis to their attorney, Vroman James Dorman. Dorman was born in 1909 in Fresno County. Dorman was a young lawyer working in San Diego trying to make a name for himself in managing finances. It was reported that Dorman and Davis did not get along. Davis was a lifelong and traditional Democrat, set on following his goals regardless of money. Dorman was a lifelong money-cautious Republican who would later serve as a delegate to the 1944 and 1952 Republican National Conventions. While Davis was a visionary who worked towards abstract ideals, Dorman was a realist and saw everything in terms of physical wealth. His only job was to manage his clients' money, so if the Andersons invested money in Davis's school, as he saw it, then it must make money.

Davis continued to live in Pacific Beach near his original campus, which was thirty miles from Carlsbad. Dorman took advantage of Davis's distance. Davis was forced to communicate by mail, as he had difficulties reaching Dorman by phone. Both Dorman's office and home phones were always busy when Davis tried calling. In a November 17, 1937 letter to Dorman, Davis wrote:

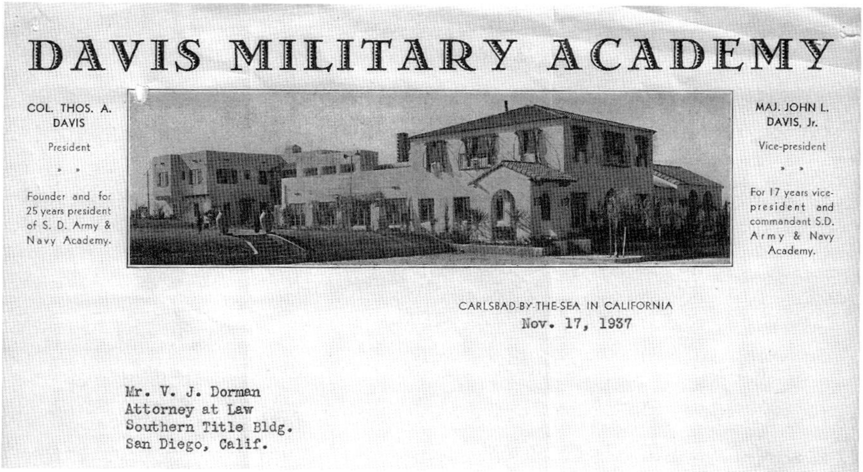

DAVIS MILITARY ACADEMY

COL. THOS. A. DAVIS
President
Founder and for 25 years president of S. D. Army & Navy Academy.

MAJ. JOHN L. DAVIS, Jr.
Vice-president
For 17 years vice-president and commandant S.D. Army & Navy Academy.

CARLSBAD-BY-THE-SEA IN CALIFORNIA
Nov. 17, 1937

Mr. V. J. Dorman
Attorney at Law
Southern Title Bldg.
San Diego, Calif.

Letter from Colonel Davis to Dorman, dated November 17, 1937, with Davis Military Academy letterhead. *Courtesy of Atkinson Museum.*

> *When in the city a few days ago, I tried to reach you by calling at your office but you were not in, and when I tried your residence, the telephone was busy….I am very anxious, in view of my friendship and appreciation of work which has been done throughout the years by our faculty members who have become interested, to get the corporation completed as soon as possible.*

The Andersons donated their own land and funded new building projects. Dorman probably saw this as reckless spending on their part. Yet even with their generous aid, the group of stockholders became discouraged and left one by one. Morale sunk, and staff began to depart. However, the biggest shock was yet to come.

During the 1938 Christmas season, Colonel Thomas A. Davis told everyone that he was resigning. This came as a great disappointment for the one-year-old Academy. Some rumors have suggested that conflicts with Dorman resulted in Davis resigning. But what surprised everyone was the fact that Davis decided to leave his position to work for the new owners of the Pacific Beach campus in the lower position of Admissions Director. This was attributed to the fact that Colonel Davis loved his old campus, the buildings he had erected "by the sweat of his own brow." Even if ownership was under another person, Davis still held the campus in the same regard as if it were one of his children.

Colonel Davis's final address to the graduating Class of 1938 perhaps foreshadowed this moment; at the very least, it summed up his feelings regarding the situation:

> *Two conflicting sets of emotions arise within my heart as each Commencement season comes around. One is the gratification I feel over the achievements you have made in your high school years, the growth and development I see in those fine bodies of yours since the first time l saw you in your school life.... On the other side, is the feeling of sorrow on your leaving for institutions of higher learning, the breaking of home ties, the loss of opportunity to serve you and to help you in the same intimate and friendly way.... So, goodbye dear fellows, I shall always prize your friendship and look back with tender memories upon the happy days gone by.*

Hostile emotions were felt by some at the San Diego Army and Navy Academy, who felt that John Brown "bought out" the original campus and stole their president. This was the spark that ignited the heated rivalry between the Army and Navy Academy and Brown Military Academy. Meanwhile, there were suspicions that Dorman had a hand in the matter, that he was the true reason Davis left Carlsbad.

Chapter 9
The Interregnum

With the departure of Colonel Thomas A. Davis, we see the end of the Academy's Davis Era. While another Davis would serve as the next president, the Academy was forever changed. Along with the closure of the campus at Pacific Beach, the following period has been described in the fiftieth-anniversary edition of *The Warrior* as part of the Academy's "Black Years." The departure of discouraged faculty members and the founder nearly signaled the end for the young campus. Until William C. Atkinson, the San Diego Army and Navy Academy would not witness a stable guiding leader. Thus, in between those two giants, between the Davis Era and Atkinson Era, we witnessed an interim period that covered the brief terms of three presidents.

John Lynch Davis Jr.

Following Colonel Davis's departure in 1938, his brother, John Lynch Davis Jr., was elected as the second president of the Academy, while Major Peterson was selected as vice-president. Major Davis led the school following his older brother's departure, reassuring faculty and staff that things were in good order. The Andersons tried to brighten the mood by financing more construction projects for the Academy.

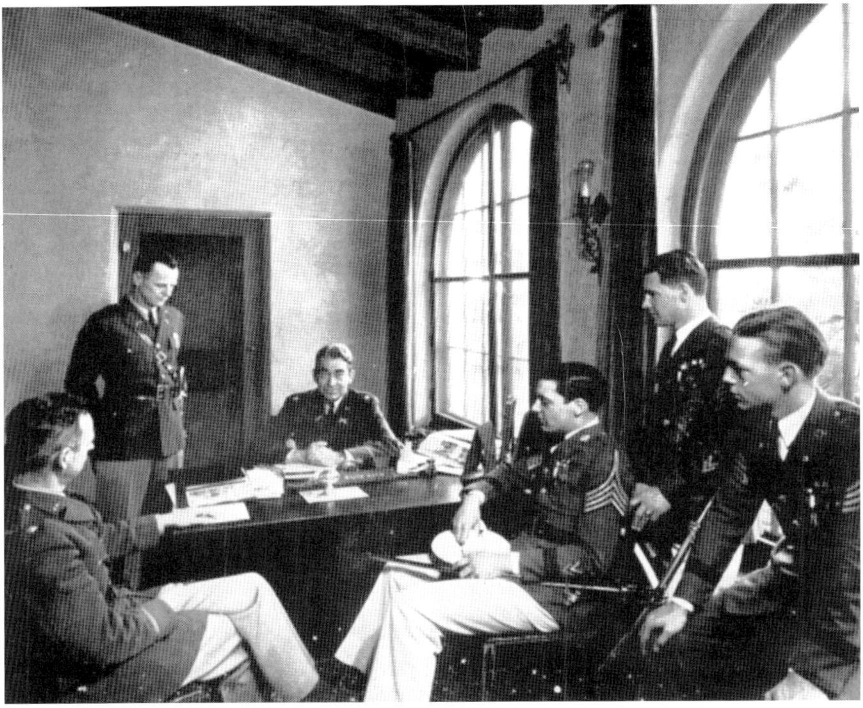

Major Davis and the Tactical Committee meeting in the mess hall. *Courtesy of Atkinson Museum.*

The Academy's football team was revived for the first time since the move to Carlsbad and, though inexperienced, competed in the Southern League.[19]

That year, the auditorium was completed. Now named Davis Hall, in honor of the departed founder, the building serves as the Academy's most recognizable landmark. The building contains the gymnasium, administration offices, classrooms, drama room, Music Department and an indoor rifle range. The Andersons also funded the construction of cottages out of their own pockets. The yearbook, *The Cadet*, was renamed *The Adjutant* that school year, while the newspaper came under the control of the Scroll and Quill Club.

In 1939, new structures were built by Alonzo "Robert" Baird to house the cadets of the new campus. Major Davis continued his brother's famous Davis System, focusing on academic and character-building studies. Like

19. While in Pacific Beach, ANA briefly played in the Southern League, which dissolved due to low numbers. The Southern League was revived around the same time the Academy revived her football team in Carlsbad.

his older brother, Major Davis was a remarkable leader in the field of education. Under his leadership, everyone felt reassured that the Academy would do just fine.

However, word of Major Davis's talents began to spread, and in 1941, he received an offer to teach at a highly respected school on the East Coast. This was a job he could not turn down, and he departed just like his brother. Records of Major Davis's life go silent after this point. This sudden departure came as another great disappointment to the Academy.

V.R. VESTAL

After Major Davis left, Captain V.R. Vestal, a graduate of the United States Military Academy, became the third president of the Academy. In certain records, he is sometimes listed as "acting president." Vestal became the first president of the "West Point of the West" who was a graduate of West Point. Very little is known of Vestal's life. He was born in 1902 and later attended West Point. Shortly after, in 1921, Vestal joined Purdue University's Cadet Corps in West Lafayette, Indiana, as the ROTC supply officer. He later joined the staff of the San Diego Army and Navy Academy. His term was less than a year, from 1941 through 1942.

Under Vestal, and by recommendation of the Andersons, the Board decided to move in the direction of turning the SDANA into a charitable, educational organization. On December 16, 1941, the Academy officially became a nonprofit organization. A new Board of Directors was created that included the Andersons, William C. Atkinson, Samuel Peterson and Vestal, in the capacity of president of the Academy. As World War II began, Captain Vestal went back into service and left the Academy.

SAMUEL WARFIELD PETERSON

Major Samuel Warfield Peterson became the fourth president of the Academy. He was known affectionately as "Major Pete" by those at the Academy. Major Peterson joined SDANA in 1916. He served as Headmaster of the High School in 1916 and then Headmaster of the entire academic program in 1929. He also filled the role of assistant vice-president.

Peterson was born in Port Huron, Michigan, on May 26, 1876. He graduated from Park College in 1899 and then continued on to the University of California for graduate school. Afterwards, he worked as a history instructor at Occidental College. He was an avid member of the Presbyterian Church.

Major Peterson was one of the five stockholders who moved with Colonel Davis to Carlsbad after the Academy was forced to vacate Pacific Beach. The Andersons, who had great admiration for Pete, advised him to dissolve the stock company since it was considered worthless. Peterson followed their advice and oversaw the reorganization.

In 1943, enrollment increased. Many saw that as a sign that the dark years were over. On the evening of Monday, April 23, 1943, the faculty threw Major Peterson a surprise birthday party. Towards the end of the party, Peterson fell ill. A doctor said that he had suffered a mild heart attack. Peterson was sent to rest in the Academy's infirmary. A few days later, he was walking around campus just fine. That Friday, April 26, the nurse knocked on Peterson's door. Receiving no answer, she let herself in. Everyone on campus was grief stricken when they found out that Major Pete had passed away in his sleep. For a quarter of a century, he was a major pillar of support, and now no one knew how the Academy would survive after him.[20] With four presidents come and gone, no one knew what kind of future the five-year-old school faced.

20. ANA's library is named in Peterson's honor.

Chapter 10
To Same Heights

It was Thanksgiving Day 1929, and William C. Atkinson and his wife were driving down Highway 101. Atkinson was a southern boy who loved making music. The two came to California to fulfill their dreams of breaking into show business. However, fate would choose an entirely different destiny for the couple.

Atkinson had just moved to California five years before. New to the area, he asked his personal secretary where he should have his Thanksgiving dinner. She recommended he visit the popular Red Apple Inn just a hop north in Carlsbad, where Margaret Drought's delicious apple pies received local fame.

As Atkinson arrived at the small Inn, he noticed the red apple theme on all furniture and light fixtures. There were even red apples on his waitress's uniform. As he sat there eating his Thanksgiving meal, little did he know that in seven years' time he would be back. As one Carlsbad history book put it, "Atkinson found his future home." After eating, he looked down at his empty plate, the image of another red apple greeting him. He would grow to hate those red apples.

William C. Atkinson

The new Carlsbad campus was just five years old, and it had already passed through four presidents. The founder and his brother left. Another left for war, while another died in office. The Academy's future was looking grim. No one knew what would become of the small school. Then, in 1943, the Academy welcomed her fifth president, Colonel William C. Atkinson. Just as Colonel Davis was the embodiment of the San Diego Army and Navy Academy at Pacific Beach, Colonel Atkinson would become synonymous with the Army and Navy Academy in Carlsbad.

Remembered affectionately as "Colonel Atty" and "Din," William C. Atkinson and Virginia Atkinson had been with the Academy since 1924. Their story, like Davis's, begins in Virginia, where the couple grew up. William Atkinson studied at Chicago Musical College, where he earned his bandmaster's certificate, and then later studied at the University of the South. William enlisted in the army for a period. Virginia studied acting and drama. They both attended the Conservatory of Music in Louisville, Kentucky, where they first started dating. William and Virginia married and then together took teaching jobs in West Virginia and Tennessee.

William worked at Greenbrier Military Academy in Lewisburg, West Virginia, from 1919 to 1921 and then at Castle Heights Military Academy in Lebanon, Tennessee, from 1921 to 1924, where he taught chemistry and physics and directed the cadet band. Virginia taught speech and drama at both schools. The Atkinsons had dreams of being in "showbiz," and that goal had brought them to California.

The Atkinsons wanted to continue teaching at a military school while still pursuing their dream. They preferred a military school in Southern California, near Hollywood, the heart of the entertainment world. In the spring of 1924, they sent inquiries to three schools. Feeling positive, they decided to place their bets on Pacific Beach.

On September 24, 1924, the Atkinsons arrived on the campus of the San Diego Army and Navy Academy. They were met with great disappointment. The Academy was in a large, undeveloped part of Pacific Beach, isolated from major communities. The campus, then only three old buildings, was in need of much repair. As they climbed the front steps, they met Major Davis, who took them to see the school and their living quarters. The Atkinsons found the younger Davis very friendly and quickly took a liking to him. Major Davis told them that the institution was still growing yet had already made a name for herself in the community; with their help, he said, the school would have a bright future ahead. The Atkinsons looked at each other and mutually agreed to take up the challenge and apply themselves with "everything they could muster."

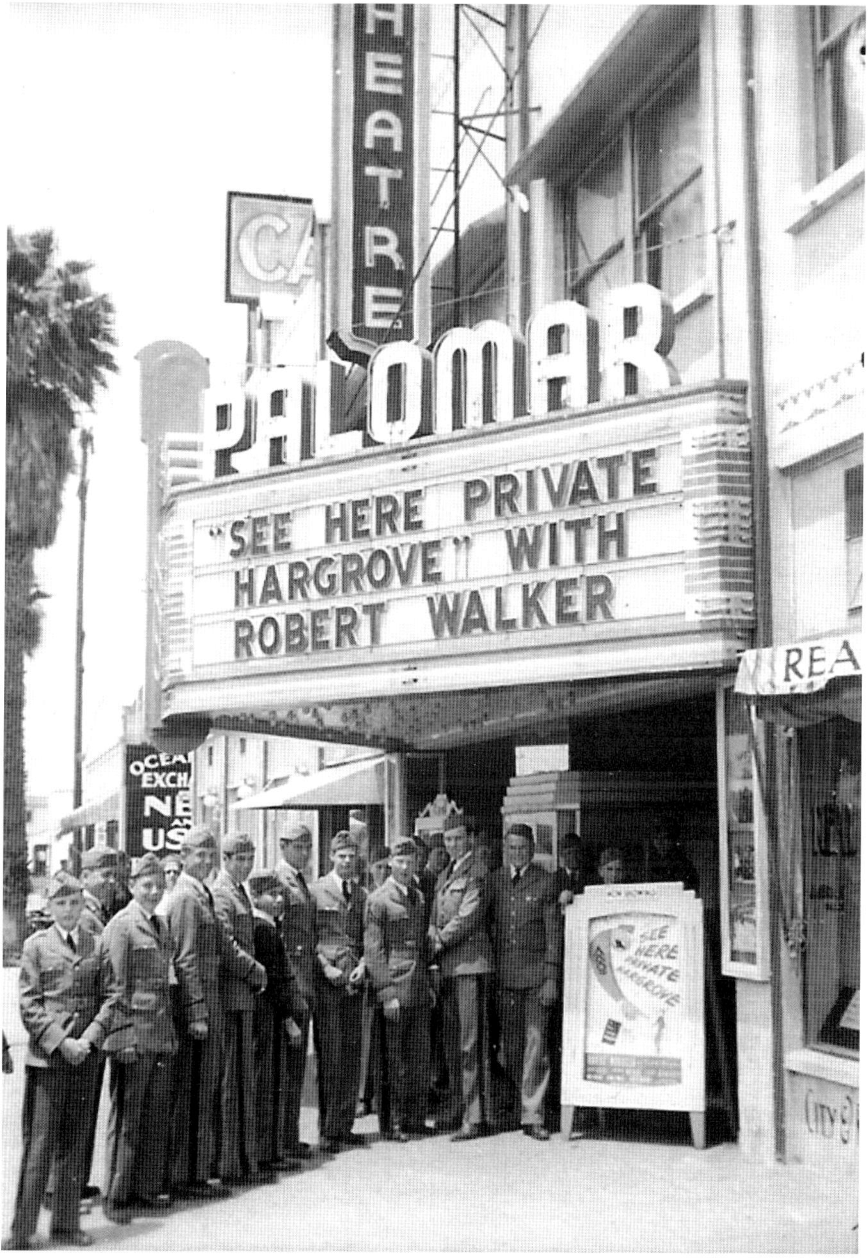

Cadets of the Army and Navy Academy waiting to see *See Here, Private Hargrove*, starring Robert Walker, at the La Palomar Theater in Carlsbad Village. *Courtesy of Atkinson Museum.*

Right away, they would prove their worth. Thanks to the Atkinsons, the Academy developed an outstanding curriculum of drama, music and public speaking. Captain Atkinson became bandmaster, and Virginia became head of the Drama and Speech Department. Captain Atkinson was soon leading the cadets in parades and performances for the local community, despite the old instruments he had to work with. In time, the Academy's Warrior Band, under Atkinson, became famous throughout Southern California. It performed on local radio stations and talk shows. In 1928, the band marched in the dedication parade for the new Los Angeles City Hall, where it won a cup for excellence in drill, discipline, appearance and military bearing. On New Year's Day, at the request of the committee in charge, the band marched in the Rose Tournament Parade for the Sixteenth Rose Bowl. In 1930, Captain Atkinson wrote the Academy's first alma mater.

On December 20, 1943, William C. Atkinson was named fifth president of Army and Navy Academy. He was promoted to the rank of colonel by the Board. His wife was named vice-president and would remain the only female vice-president of a military academy until the modern era.

Virginia Atkinson

Tantamount to the extraordinary job her husband had done with the marching band, Virginia Atkinson made a lasting impact on the Academy's Drama Department. In 1924, she formed the Masque and Wig Club. The motto of the club was "*Ich Dieu*," meaning "I Serve." The Academy's Masque and Wig Club grew in fame, as it went on to win numerous competitions across the nation. The club was arguably one of the best High School drama clubs during this period.

In 1945, the Masque and Wig received a Thespian Charter. The club consecutively placed first for years in the California State One Act Play Tournament, sponsored by the Pasadena Playhouse. The Playhouse awarded the club special honor for production, direction and performance. In 1951, Fredric Berest, actor in films such as *Voodoo Tiger* and *Serpent of the Nile*, wrote the Academy a letter requesting she not enter the Pasadena Playhouse contest again by virtue of the fact that there was "no competition," as the Masque and Wig always won the Governor Earl Warren Trophy. The cadets of Masque and Wig became so popular that they were requested to perform across the nation.

Mrs. Atkinson was beloved not only by the cadets but also by the children of the local community. Joe Orozco, who worked at the Academy's mess hall for half a century, fondly remembered her generosity. As he recalled, growing up in Carlsbad meant there was not a lot to go around. It was a very poor city still recovering from the Depression. However, every year the children of Carlsbad would get to enjoy the Christmas party put on by Mrs. Atkinson, who would host plays for the entire city during the holiday season at the Carlsbad Theater and Davis Hall. Joe's first memory of Mrs. Atkinson was when she handed him a Christmas stocking full of goodies as a child.

Virginia Atkinson left a strong impact on everyone who was fortunate enough to cross paths with her. One cadet who credited her with changing his life was actor Robert Hudson Walker. Robert "Bob" Walker was born in Salt Lake City, Utah, on October 13, 1918. He was the youngest of four children and, as a child, was emotionally scarred by the divorce of his parents. Bob became a troubled child unsure what to do with life. In 1932, he was sponsored by his aunt, who put up the funds to send him to the San Diego Army and Navy Academy while the campus was still in Pacific Beach. As Walker later described his experience, "Although I didn't like discipline, it's life for you that I should wind up at the San Diego Army and Navy Academy, where discipline IS discipline!"[21]

At the age of thirteen, young Bob was miserable at the Academy. Mrs. Atkinson saw Bob alone and took to the boy. She noticed that he had excellent rhythm playing the drums and asked him to read a script from a play. As Walker described it:

> *I had been at the military academy about a year, and hating every part of it, when one day Mrs. Atkinson asked me to read. She was a kind and wonderful woman, and she must have been a really fine psychologist, too, because she got me so interested in dramatics I forgot how much I hated discipline. For the first time in my life, my future came into focus. From that day to this one, I have had just one ambition—to be a good actor. What would have happened if she hadn't seen something in me…I don't know. I think that perhaps my whole life might have been changed. Of course, it helped that I thought she was the loveliest woman in the whole world!*

After the reading, Mrs. Atkinson asked, "Robert, how would you like to be in the next school play?" Walker was shocked; he never acted

21. Information comes from newspaper articles archived at the website Robert Walker Tribute.

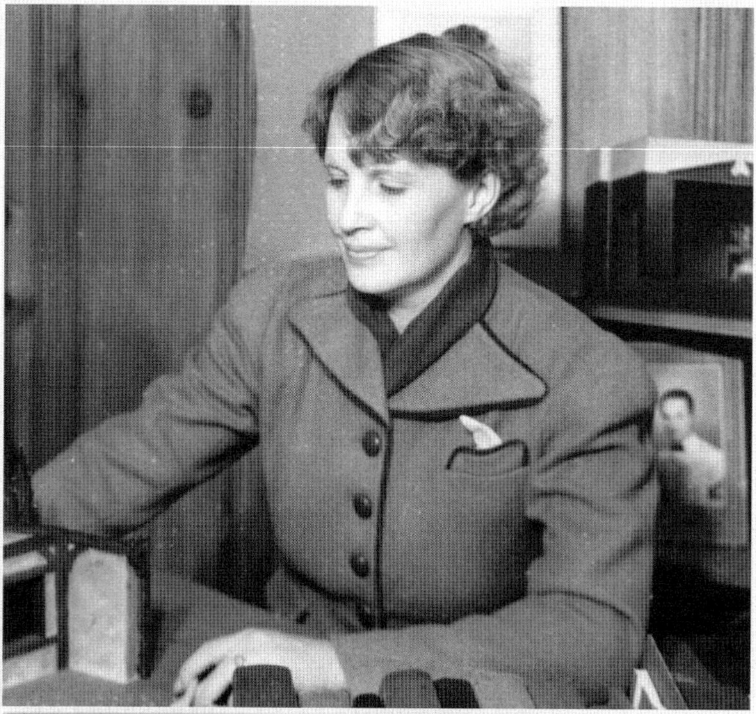

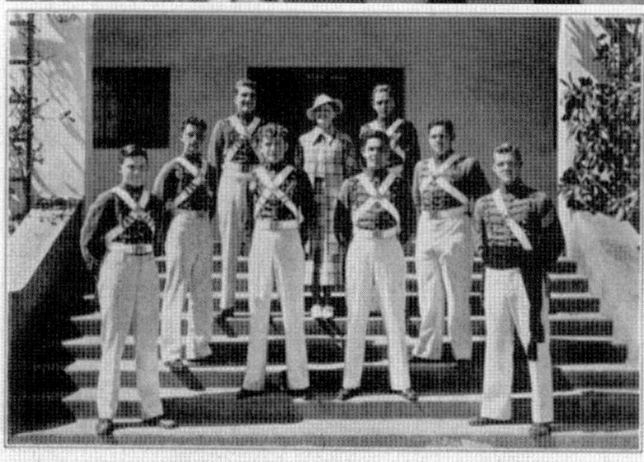

Left to right: Warren, F., Gorsun, Horton, Walker, Mrs. W. C. Atkinson, Hull, H. F., Buckley, Dryden, Throneson.

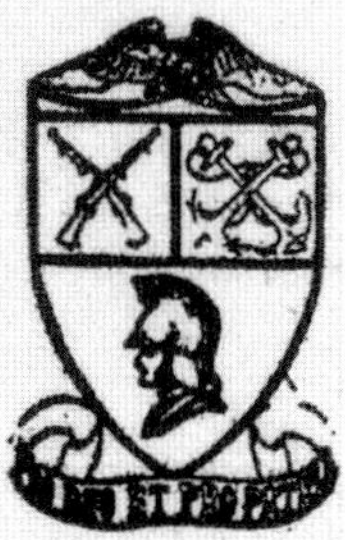

Top: Atkinson, working at her desk as vice-president of the Army and Navy Academy. *Bottom*: The 1936 drama club. Virginia Powell Atkinson is in the center back row, and Robert Walker is the cadet in front of Atkinson to the left. *Courtesy of Atkinson Museum.*

before in his life but decided to take up the offer, as it was an alternative to getting in trouble.

Later that year, Walker won first place in the district competition at the Pasadena Playhouse. In 1935, Walker won "Best Actor in San Diego County" for his performance in *The Other Side*. That night, he wrote to his family, "Dear Mother, I guess tonight I am the best young actor in all San Diego."

In 1936, he won the Best Actor award again and was offered a scholarship to the Pasadena Playhouse. When the Academy moved to Carlsbad, Walker followed Mrs. Atkinson to the new campus. Walker graduated with the Class of 1937, the first class to graduate from the new campus. Afterwards, he enrolled in the American Academy of Dramatic Arts in New York City.

Walker mentioned that while he was not interested in girls at that age, he did "immediately became conscious of the most beautiful woman in the world—Mrs. Virginia Atkinson, my dramatics teacher." For the rest of his life, he would mention meeting Mrs. Atkinson as his "first break," according to Canfield's "You Can't Do That, Bob Walker." Until he passed, Walker always kept in touch with his drama teacher, hoping that she would be proud of him.

Robert Walker returned to Carlsbad in 1944, when *See Here, Private Hargrove* was filmed in the city and Camp Pendleton. He played the lead role and stayed with the film crew at a hotel a few blocks from the Academy. Walker was pleased to see his old drama teacher once again. He would go on to star in Alfred Hitchcock's *Strangers on a Train* in 1951, his most memorable role. Sadly, he died before production could finish. One month later, his mentor passed away.

THE WAR YEARS

On November 30, 1943, the Andersons stepped down from the San Diego Army and Navy Academy's Board of Directors. They continued to aid the Academy in her standing as a nonprofit organization in order to receive federal tax-exempted benefits. The Andersons suggested that the Academy cease renting from them and enter a purchasing program that would extend over a twenty-year period. The Academy (including property, real estate, buildings and equipment) was estimated to have had a value of $225,000 at the time.

The Andersons arranged to transfer the title to the school, calling for twenty annual payments of $10,000, which would completely pay off the mortgage. Dorman oversaw the process and drew up the papers to convert the Academy from a private corporation to a nonprofit educational organization.

The Board of Directors was reorganized with five members: Colonel William C. Atkinson as president of the Academy, Virginia Atkinson as vice-president, Vroman J. Dorman, O.M. Morris and Fred W. Mitchell. The size of the Board was allowed to increase, but at the discretion of the majority. In 1943, "San Diego" was dropped from the name, and the school was thereafter known as the Army and Navy Academy. As one advertisement put it, from the "Academy's perspective":

> *In 1943 I became incorporated as a nonprofit school, and, at the time, I dropped "San Diego" from my name. If, for any reason, I should die, my property would go to the State of California. I, then, own my own campus with no person having a nickel vested within it.... To guide me in the right direction, I have a board of directors consisting of five men with three of them having been my former students. In guiding me in the right direction, these men insist that I ever remain an old-fashioned school.*

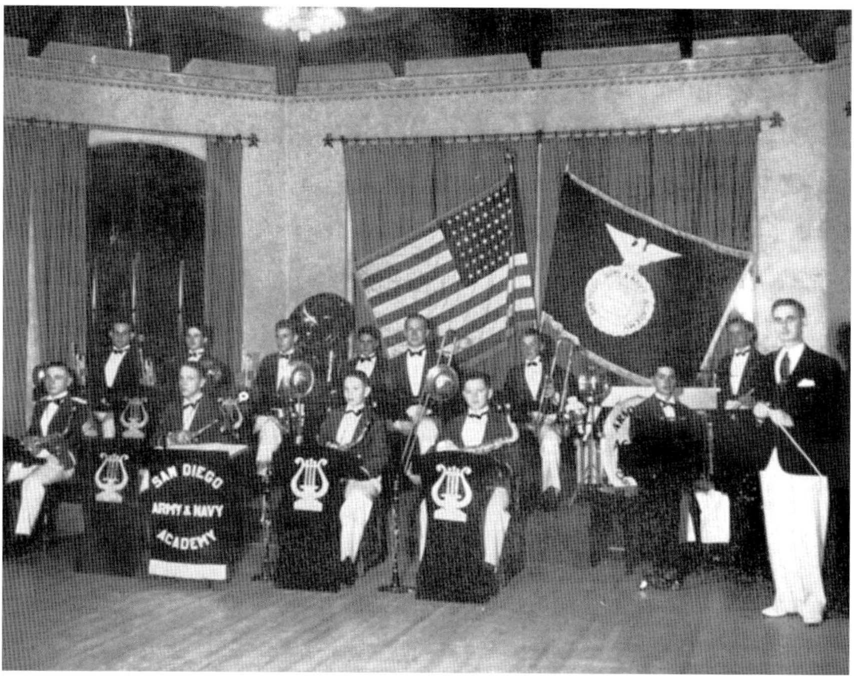

Bandmaster Atkinson and the band, shortly before the school dropped "San Diego" from its name. *Author's collection.*

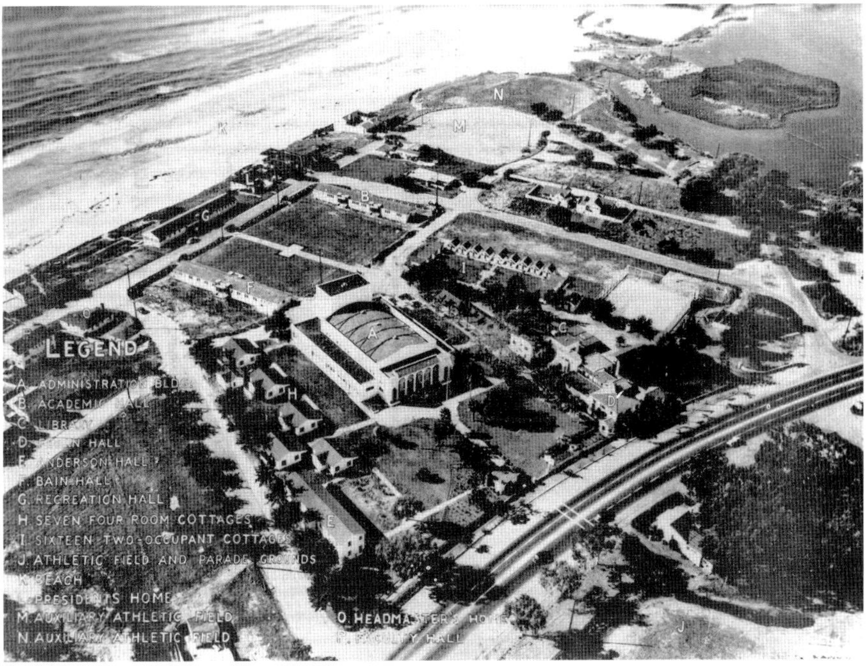

An aerial view of the 1940s campus, with the newly built two man cottages making up North Court, seen top right of Davis Hall in the center. *Author's collection.*

That same year, Colonel Atkinson established the Silver A, the highest honor awarded to a cadet at the Army and Navy Academy. A recipient had to be "a young man of good character, set a good example and demonstrate the utmost responsibility and trustworthiness" and "look out for his fellow cadets and demonstrate loyalty to ANA and his cadets." The award has been noted as the "epitome of what the Academy strives to achieve." The Silver A is only awarded to a select few senior cadets; sometimes it is not awarded at all if there is no deserving candidate. The first Silver A recipient was Bruce Myers, Class of 1943. Recipients of the Silver A normally went on to have successful futures, including California senator David Kelly (1948), General William Crouch (1959) and *The Amazing Race* winner Uchenna Agu (1982).

In December 1943, in remembrance of Pearl Harbor, the senior class planted a tree on the lawn opposite the Senior Lawn. Since then, the tradition of tree planting has continued every year on Pearl Harbor Day.

As the Academy's popularity regained traction, more young men wanted to enroll. By 1943, enrollment was up to 175 cadets. When Mr. Anderson heard that boys were being turned away due to limited space, he spoke to

Colonel Atkinson about the matter. Atkinson told him that the Academy was completely filled and that they could not financially build additional dorms. The Atkinsons vowed never to ask the Andersons for aid; however, the Andersons voluntarily supported the construction of more dorms. The Andersons spoke to Atkinson's architect and told him that they would take care of all the bills. In the meantime, tents were used to temporarily satisfy the increasing enrollment. While increased enrollment was viewed as a good sign to many, Colonel Atkinson stated that his aim was not to strive for larger enrollment but rather to concentrate on the quality of complete education and more intimate devotion to fewer cadets.

In 1944, Anderson Hall was completed. This was the first permanent dormitory on the new campus and named in honor of the Andersons. Anderson Hall was also the first building constructed in Carlsbad under the War Production Board regulations; the war effort made building materials difficult to obtain. To construct the building, Atkinson had to appeal directly to the main office of the War Production Board in Washington.

The following summer, sixteen two-man cottages were constructed in the place of the tents. These buildings, collectively named North Court, were temporary buildings meant to be replaced. To this day, only one of these cottages remains, used as a storage shed.

CAMP PACIFIC

In 1943, Camp Pacific, the Academy's summer program, was established. Originally opened as a surf camp, presently it functions as a multipurpose summer camp. Camp Pacific itself is a co-ed, nonmilitary summer camp program containing the Academy by the Sea, which operates as the academic portion, along with a surf camp, leadership training camp and a wide variety of other types of educational and summer camps. In 1995, the Academic Summer Camps were introduced as Camp Brain to provide students with many academic and enrichment classes.

In 2010, the Summer Leadership Programs were added. These programs continued the Academy's leadership education program during the summer, when Academy cadets and co-ed students from other high schools could develop their leadership and team-building skills. This program is a great advantage to cadets from other JROTC units for the extra experience and training they would not receive at their own schools.

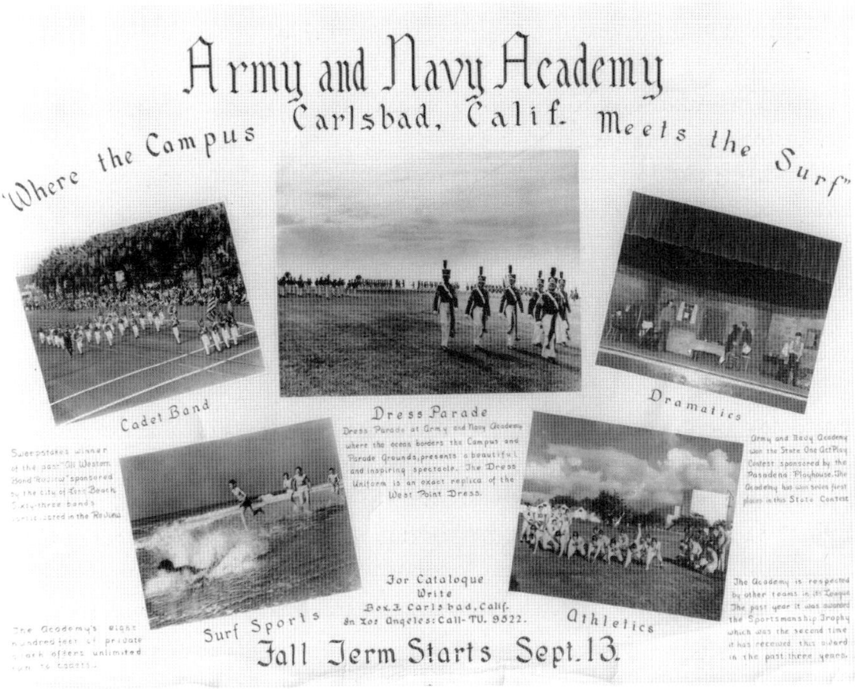

Original advertisement with handwritten ink on photograph, denoting the Academy as the place "where the campus meets the surf." *Author's collection.*

Camp Challenge is run especially for the JROTC program and is the only JROTC summer credit program west of the Mississippi. The Academy also runs Camp Challenge Jr. to give Middle School students the same experience as the High School cadets.

Annually, Camp Pacific also welcomes a wide array of other camp programs, offering space to many other educational organizations. Today, Camp Pacific is the largest summer camp program west of the Mississippi.

OLD FAMILIAR FACES

Colonel Atkinson had a knack for hiring great instructors. He was even able to convince his former coworkers from Pacific Beach to leave Brown Military Academy to reunite with their old friends in Carlsbad. Such familiar faces included Captain Charles Bain, Major James Moir, Captain Edger Mohan, Captain Raymond Ede and Band Director Al Polhamus.

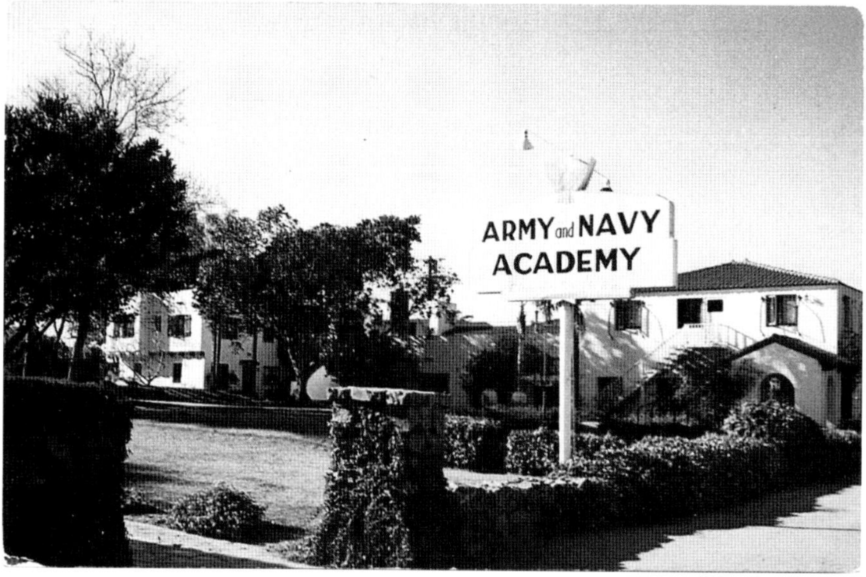

Sign of the newly renamed Army and Navy Academy. *Author's collection.*

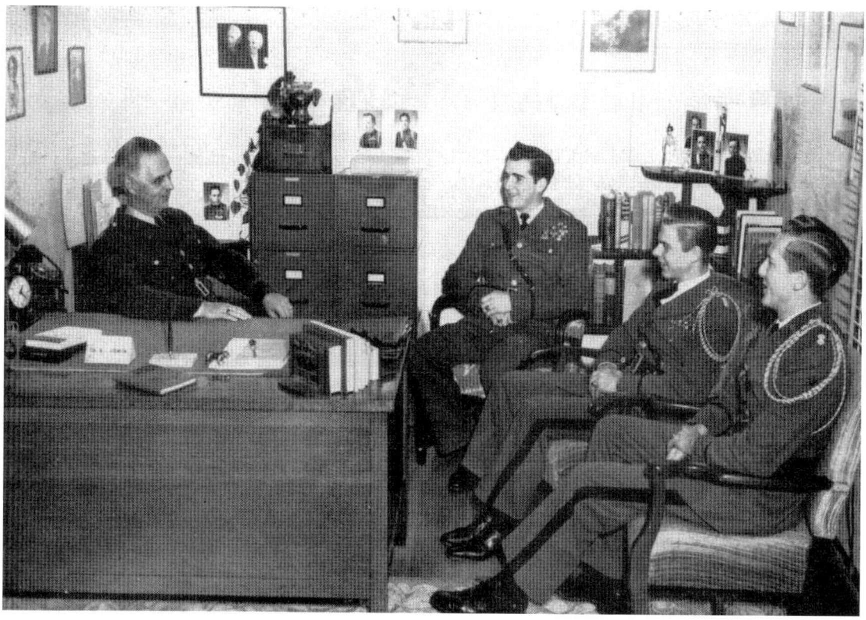

Colonel Atkinson in a conference meeting with members of the Class of 1951. *Author's collection.*

James Moir was a captain in the British army who had worked at the Academy in the Science Department since 1926. He worked as director of admissions for Brown before returning to Army and Navy. Edgar Mohan, a graduate of the University of Illinois, worked as Headmaster of the Junior School since 1924. The building behind the mess hall is named in his honor. Al Polhamus served as bandmaster under Atkinson. The two bandmasters would go on to write the current Army and Navy Academy alma mater, along with numerous marches and fight songs.

In 1947, Raymond Clark retired from his job as the athletic director. Henry "Coach" F. Lawrence took over the position. Coach Lawrence came to the Academy as science instructor but became known for coaching the football team. His motto was "Carry a big stick and yell."

In 1948, the Academy expanded with three new buildings: the Recreation Hall, the Academic Hall and the new dormitory, Bain Hall, named in honor of the late Headmaster. These buildings had belonged to the army during the war and were bought by the Academy and moved to the campus in sections.

Around this time, the first Commanding General of the Marine Corps Base Camp Pendleton, Major General Joseph Fegan, a loyal friend of the Army and Navy Academy, was building a lasting bond between the Academy and the Marine Corps base. General Fegan had given the Academy *The Truck*, an arrow ornament that was placed on top of the flagpole. He explained that the arrow pointed towards Camp Pendleton, signifying a special friendship between the Marine Corps base and the Academy. Sadly, half a century later, during the Centennial celebrations, *The Truck* was removed from the original flagpole and replaced with an inapt yacht yardarm. In 1948, the Red Apple Inn was renamed Fegan Hall in the general's honor.

On April 28, 1949, Colonel Atkinson established the Honor Council, a peer-enforced conduct system similar to a military court-martial. The group is made up of the top senior cadets in the Corps whose job it is to uphold the Academy's Honor Code. If a cadet were caught violating the Honor Code in any way, he would have to face a trial to determine his guilt and punishment. The Honor Council at the Army and Navy Academy is notable in that aside from a faculty advisor, all investigations and hearings were solely determined by the cadets. If guilty, the proper punishment would be decided by the council and recommended to the Academy's Commandant or Headmaster. This gave these cadets a greater burden of responsibility and good experience for testing their leadership education.

PART IV

THE ATKINSON ERA

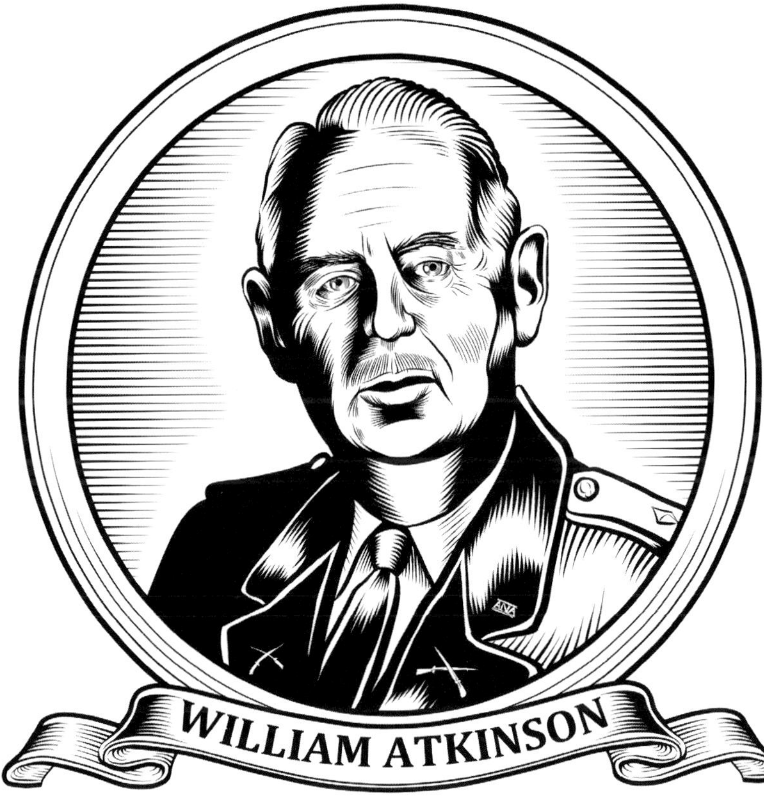

William C. Atkinson. Artwork by Alexander Mui.

Chapter 11
Atkinson's Lament

Colonel Atkinson shed tears as the Army and Navy Academy held a memorial service for the Andersons. In August 1949, Arthur C. Anderson passed away. Eight months later, succumbing to illness, Gertrude H. Anderson passed away. They were not just friends of the Academy—they were his dear personal friends. Atkinson learned that the Andersons left all their property to the Army and Navy Academy, debt free. Their wish was that the school remained nonprofit and nonsectarian and that the property always be used for the education of children.

Dorman helped Atkinson manage the transition after the Andersons' passing. Initially, Atkinson was wary of Dorman; he knew that Dorman was a major factor for Davis deciding to leave the Academy. Dorman was a realist, always money cautious. Atkinson was an idealist who stuck to doing things the old-fashioned way. A perfect example, as alumni recall, was that Atkinson would have the same speeches saved every year for each week's chapel service. However, the two would balance out and keep each other in check.

In the fall of 1951, enrollment was up two hundred cadets. That year, the Academy newspaper was renamed *The Warrior*. Things were looking up until Atkinson and the Academy dealt with a harsh misfortune. On Christmas Day, the vice-president of the Army and Navy Academy, Virginia Powell Atkinson, passed away after serving the Academy for twenty-seven years. For the next few months, Colonel Atkinson mourned her loss, making daily trips to the cemetery. The 1952 yearbook was dedicated to Mrs. Atkinson:

To the memory of Virginia Powell Atkinson, a gracious lady, a loyal friend and an able teacher who spent 27 years of her life as a teacher, drama director and vice president, and who offered guidance, kindness and help to every Cadet who sought it. Her memory will live ever alive and bright in the hearts of those who knew her and loved her.

THE ACADEMY AND CARLSBAD

After the departure of Colonel Davis, it was Atkinson who brought the Army and Navy Academy back to prominence and success. However strong a leader he was, Colonel Atkinson was still human, and the loss of his dear wife and the Andersons came to him as a serious blow. He was determined not to experience such defeat again. Now alone, he would pour his heart into the Academy, in dedication of his late wife and friends.

Colonel Atkinson set out to improve the Academy; he was concerned not with enrollment numbers but rather quality of education and cadet care. However, by now it had become evident that Carlsbad was perhaps not the best location for the school during this period. It still had not completely recovered from the Great Depression nearly two decades ago and was seen as a "financial dump." The mineral water, spas and hotels just weren't bringing in the money they had in the town's heyday. During the Cold War, such luxury and extravagant recreation seemed hollow in the midst of fear of global nuclear annihilation.

Keeping the cadets a top priority, Atkinson became increasingly disappointed with the state of the area. The city had no water pipes, local hospitals, fire departments or police stations. Carlsbad was not incorporated and thus did not have the infrastructure of nearby prosperous cities like Oceanside to the north. If a fire broke out, if a cadet were injured or if a crime occurred on campus, the Academy would be completely isolated and helpless. He hoped that the town would modernize quickly.

From Atkinson's perspective, it was the Army and Navy Academy that saved the small town and "did much to place Carlsbad on the international map." Through promotional advertisements sent to foreign markets, he believed that the Academy did "more to bring Carlsbad to the attention of the international public than any other agency" in the area, counting the monthly family visits, parades and weekend events that brought "business to motels, restaurants, service stations and other businesses in general." With the decline in revenue that never recovered following the Great Depression, it is fair to say that ANA was one of the big profit makers for the surrounding community. From that perspective, the colonel's request to the city to improve its infrastructure was not out of line, as modernization would be beneficial not just for the Academy but for the entire community.

In October 1951, the Academy and Warrior Band helped dedicate the new Carlsbad Post Office, one of the first modern buildings the city constructed following World War II. For Atkinson, this was a sign that Carlsbad was slowly making the improvements it needed, and he hoped that conditions would continue to improve.

On New Year's Day 1952, shortly after the passing of Virginia Atkinson, a fire broke out in Carlsbad. As Colonel Atkinson feared, the city was not prepared to deal with this chaos. The Carlsbad Volunteer Fire Department had no truck, and the citizens of Carlsbad had to stand by and wait for outside help. A nearby forestry truck arrived only to have its pump malfunction.

Half an hour later, when a fire truck from Oceanside and one from Del Mar arrived, the firefighters noticed that there was no hydrant in the vicinity

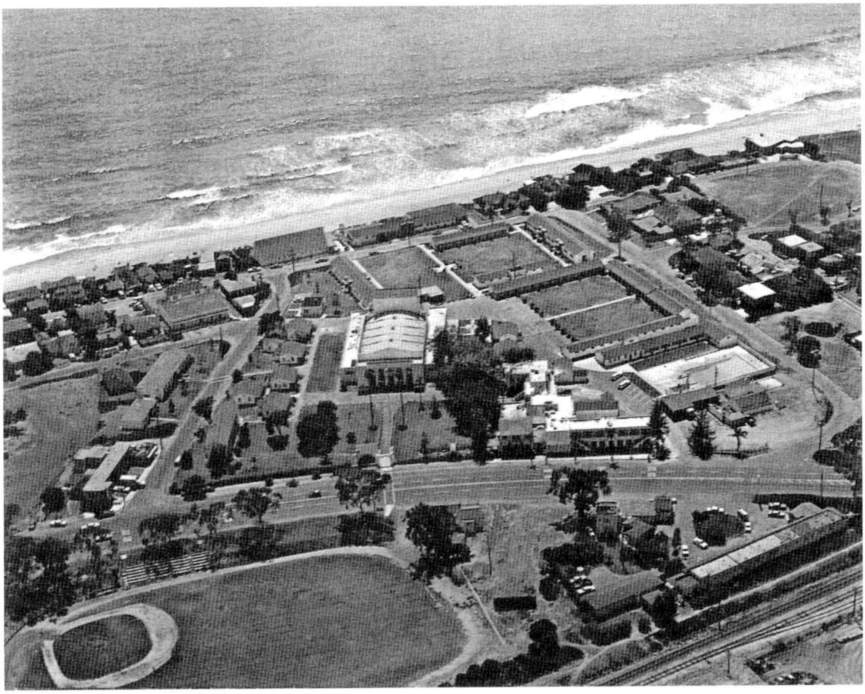

An aerial view of campus, circa 1950s. Everything above Carlsbad Boulevard running down the lower two-thirds of the photo was part of the Strip. *Author's collection.*

of the fire. They were forced to keep refilling their tanks at Pine Avenue School, rushing back and forth between the hydrant and the fire. After three hours of snail-paced trips to the hydrant, the fire was finally put out.

Major General W. Worton, a retired Marine and former Los Angeles police chief, lived next to the building destroyed by the fire. He noticed that the rubble was still smoldering. An hour after the fire was thought to have been put out, flames erupted again. Worton called for the forestry truck; however, the truck was now in La Mesa, attending to other matters. It took an hour before a truck from Del Mar arrived. It took another four hours before the fire was contained.

Major General Worton was especially vocal with the manner in which the situation was handled. He criticized the officials of Carlsbad while writing a letter of thanks to the Oceanside mayor. He was not alone; numerous residents voiced their complaints about the lack of responsibility from officials. The editor of the *Carlsbad Journal* wrote that Carlsbad was "growing like Tospy without guidance." Other issues were also brought up that painted a picture of Carlsbad's deteriorating state, including complaints of a "lake sized" puddle forming on Oak Avenue after every rain. The supervisors responded by telling the complaining residents that they would be provided with sponges for the puddle.

THE ANNEXATION COMMITTEE

This disaster was the last straw for Colonel Atkinson, who had just spent the holiday mourning his wife's passing. He was fed up with the outdated conditions of Carlsbad and demanded action. Atkinson did what any southerner fed up would do: he decided to secede.

On February 22, Colonel Atkinson gathered the group of hotel owners and landowners west of Carlsbad Boulevard and formed the Annexation Committee. The committee's intention was to have the land and property annexed by Oceanside, which had running water from Colorado Springs, medical hospitals and operational firehouses. Colonel Atkinson was named the chairman of the Annexation Committee, and the group dedicated itself to carrying out the plan. The members approached Oceanside with the proposal. General Worton was an avid supporter of annexation, as the fire had broken out near his home and he owned most of the beach property nearby. By this point, he gathered assurances from Oceanside officials and members of the Oceanside City Council that they strongly expressed interest in gaining the additional land.

This land west of Carlsbad Boulevard became known as the "Strip" in the *Carlsbad Journal*. The Strip contained the Academy's property, including athletic fields and the Point, which was owned by the Academy at that time. This land also contained the beachfront houses of wealthy residents, landmark hotels and numerous highly valued properties.

The Annexation Committee worked with Oceanside officials to ensure that they would be taxed fairly, receive representation on city committees and have a fire station built in south Oceanside one minute from the Strip.

On March 11, Colonel Atkinson and the committee obtained fifty-seven signatures of the ninety registered voters of the Strip and presented their petition to the Oceanside City Council. Then came the biggest blow to Carlsbad: it was announced that *all* land west of the freeway down to Batiquitos Lagoon would be part of the annexation.

THE INCORPORATIONISTS

Raymond C. Ede began working with the San Diego Army and Navy Academy in 1929. He taught English and French to cadets. When Colonel Davis moved to Carlsbad, he remained in Pacific Beach. Charles Bain, the Headmaster, also stayed. Charles Bain had worked with the Academy since 1924, when he joined as an instructor in the Science Department. Bain graduated from the Royal University of Ireland and University of London, where he obtained a degree in chemistry. He served as headmaster at Tate School in Wexford from 1907 to 1920 and then as headmaster of Cavan School from 1920 to 1924. At the Academy, Bain went on to serve as director of the summer program, Junior School Headmaster and then Upper School Headmaster.

Charles Bain had a daughter, Dorothy, who lived with him at Pacific Beach. Ede was very close to the Bain family, and when they remained at the old campus after SDANA moved, he decided to do the same. Ede and Dorothy eventually married. The Pacific Beach campus was renamed Brown Military Academy after being purchased by John Brown. They were later rejoined by their old boss, Colonel Davis, who returned to work as assistant Admissions Director under John Brown.

Brown Military Academy became the Army and Navy Academy's biggest rival in sports and military drill competitions. The staff of the Carlsbad

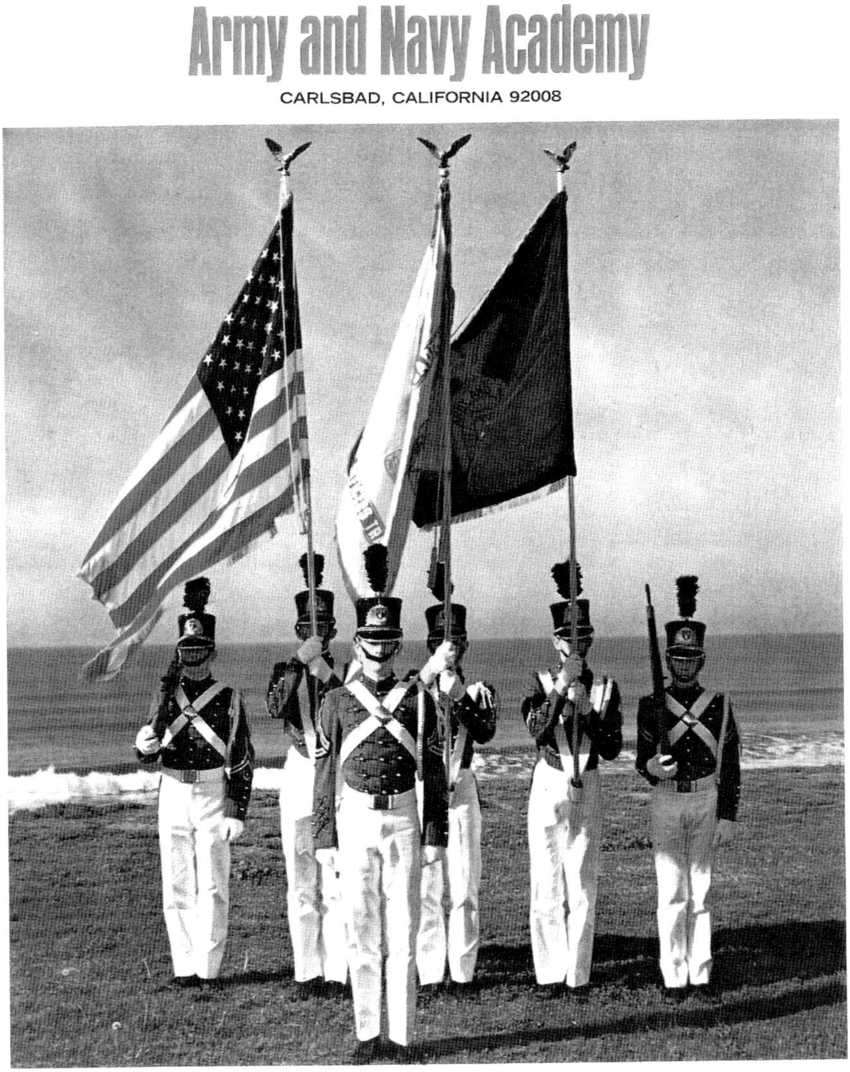

Cover of the 1954 catalogue. *Author's collection.*

campus had bitter feelings towards Brown, which they saw as an occupying force on the old campus. When Brown "stole" Colonel Davis, it was like rubbing salt in the wound. This situation must have made games between the ANA Warriors and Brown Warriors awkward for former friends and coworkers who cheered from opposite sides of the field. However, campus operations under Brown were not the same. The new administration was at the whims of the Board of Brown School all the way over in Arkansas. As they were so far removed, any issues the school faced were distant concerns, as focus was put on the flagship school, John Brown University. However, the Brown Board sent its own employees (whose main concerns were evangelical preaching and enforcing a religious structure) to fill in leadership roles, pushing longtime Pacific Beach faculty to lower positions. Many were dissatisfied under this new administration. When Colonel Atkinson reached out to his old friends remaining at Pacific Beach, all left for Carlsbad after seeing the Army and Navy Academy's success under Atkinson's guidance.

Ede came to Carlsbad with the Bain family in the early 1940s. In 1942, Bain continued his role of Headmaster at the new campus. When Bain passed away in 1945, Ede took over the role from his father-in-law. As he married Dorothy, the title of Headmaster now remained in the family. In time, Raymond Ede would be remembered as the Academy's greatest Headmaster. He would serve in that role for twenty-nine years, the longest of any Headmaster.

While a new resident to Carlsbad, Ede quickly grew to love his new home. He became heavily involved with the local community and prayed at the historical St. Michaels's Episcopal Church. While it was true that Carlsbad needed many improvements, Ede saw that the only way for Carlsbad to grow was through incorporation. Ede was a resident of the Strip, and while he worked well with Atkinson, he did not share his views regarding annexation. In fact, Ede felt just the opposite. Ede was one of the leaders of a group of citizens that favored the incorporation of Carlsbad, having been working towards it for some time.

When it was announced on March 11 that Atkinson's committee sought to annex *all* land west of the freeway to Batiquitos Lagoon, Ede could not stand idle. This land literally included all meaningful areas of Carlsbad in terms of tourism and taxation. Not merely the Strip, this included the historic founding well, hotels, mineral spas, the resorts, all beachfront property, residences of the wealthy, important commercial and industrial property, Terramar and the Encina power plant. If Atkinson succeeded with his plan, Carlsbad's hope for incorporation would vanish, and there would be no possible way to support a new municipal government. The *Carlsbad Journal* ran the headline, "Carlsbad May Lose $1,000,000 Assessed Valuation."

Atkinson versus Ede

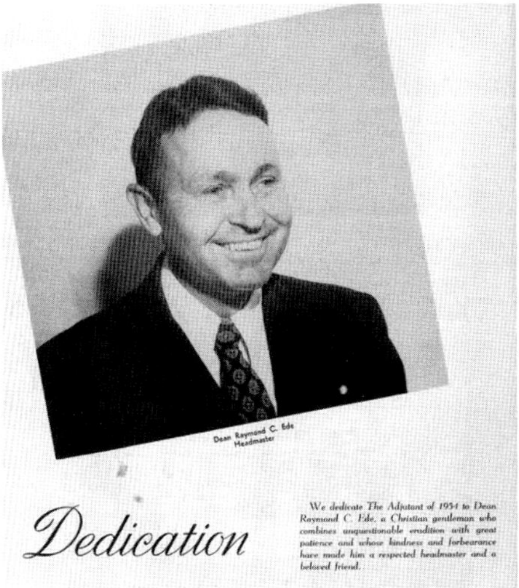

A page from the 1954 yearbook, dedicated to Headmaster Raymond Ede, one of the leading founding fathers of an incorporated Carlsbad. *Author's collection.*

Civil war erupted on campus. The Academy's president led the Annexation Committee, while the Headmaster helped spearhead the campaign for incorporation of Carlsbad and led the charge against annexation. It is not known how Atkinson and Ede felt about each other during this period, but it is safe to speculate that tensions were high. Atkinson, who still mourned the recent loss of Virginia, would not back down, and neither would Ede.

The Annexation Committee launched the first attack on the matter by placing a full-page ad in the paper noting the cost benefits of joining Oceanside. That began a war between the "Incorporationists" and the Annexation Committee, fought through ads in the local newspapers. The ads would respond to the opponent's previous ad in a chain that continued for many weeks, each refuting the other. Finally, when the Incorporationists took out an ad pointing out that Oceanside's assessed valuation rate was 20 percent higher than the county figures, the Annexation Committee ceased running ads.

On March 26, the Oceanside City Council held a public hearing on annexation. The chamber of the building was overflowing, and emotions were high. May 2 was chosen as the date for the election to decide the matter. The Incorporationists met with a crowd of more than three hundred people two days later. On April 8, both sides met the board of supervisors. Oceanside's city attorney said of the Incorporationists, "We are not here to be impressed by mobs."

Voting took place on May 2 in the Carlsbad Hotel. The count held forty-four in favor of annexation and forty-one opposed. Incorporationists argued that five of the votes did not count, as two of the votes came from residents from Oceanside and the other two were hotel employees who did not even live in Carlsbad. And there were also five more absentee ballots to count.

The recount took place on May 14. The final result was forty-five for and forty-five against. The Annexation Committee lost to a tie vote.

After the failed annexation, a group of Carlsbad citizens met at the Twin Inns to carry out its goal of incorporating the city. Raymond Ede helped lead the charge. On July 16, 1952, Carlsbad became an incorporated city. C.D. McClellan served as the city's first mayor, while Ede served as one of the first city officials. In 1955, Ede became the second mayor of Carlsbad, while Colonel Atkinson returned his focus to the Academy.

CHAPTER 12

TOWARDS THE SUNSET

COLONEL DAVIS RETURNED TO THE CAMPUS he built and to which he dedicated his life. The name "Brown Military Academy" was plastered everywhere. As with his school in Pasadena, it was as if his legacy had been completely erased. The catalogue for Brown Military Academy noted that it was the "only honor military school" in California, an honor technically earned by the San Diego Army and Navy Academy, now credited to another man. Davis was once the most respected leader in the military school community, but now he found himself willing to exchange his role as president of his Carlsbad school for any job Brown would provide.

As Davis walked through his old home, he noticed John Brown's presence everywhere. Brown was a respected educator. He owned numerous institutions across the nation and headed an entire school system that bore his name. Before him, Davis saw his lifelong dream achieved by another individual.

Davis returned to his old office and submitted his application for employment. His position of president was already occupied, as were the vice-president, Headmaster and Superintendent positions. So, what job would the school be able to offer its founder?

Near the end of the faculty section of the 1939 yearbook, there is a page with a small picture of Colonel Davis in the bottom corner. The expression he wears is that of a broken man. The photo's caption asked Brown Military Academy to welcome its newest staff member, the new *assistant* Admissions Director.

John Edward Brown

John E. Brown was born in 1879 in Arkansas. As his family faced financial difficulties, Brown was not able to receive a proper education and was forced to work at the age of eleven. He spent his childhood as a laborer in Arkansas until he encountered the Salvation Army. This experience changed his life, as shortly after he became a Methodist evangelist and began traveling the Midwest preaching the gospel.

Brown gained some fame as a young evangelist and publisher, enough that he was offered the position of president at Scarrit Collegiate Institute in Neosho, Missouri. Thus, at the age of twenty-two, the man with no traditional education became the youngest college president in the nation. During this period, he fancied opening his own college, one dedicated to educating youth with traditional Christian values.

In August 1919, Brown, with help from his friends and family, worked towards creating a new school. The cornfield from his family's farm in northwest Arkansas, along with their home and printing plant, was used for the grounds of the new school. It is not known how a boy from a poor family hit upon the financial means to fund such an endeavor, but nonetheless, one month later, on September 30, the Southwestern Collegiate Institute opened its doors. The academy and junior college opened with twelve teachers and seventy students. In the spring of 1920, the school was renamed John E. Brown College.

In the fall of 1925, Brown purchased the entire community of Sulphur Springs and opened John Brown University. Eventually, he would come to open many more schools, such as the Siloam School of the Bible, John Brown Vocational College, Julia A. Brown School for Children and the John Brown College and Academy for Female Students, all under the banner of the Brown College Corporation. Brown also bought many radio stations to spread the word of God over the airwaves, including KFPW of Missouri, KUOA of Arkansas, KGER of Long Beach and Rod of God station in Siloam Springs. Brown also owned the John Brown University hospital, was editor of two Christian magazines and was president of American Federation of Christian Workers.

During the Great Depression, the John Brown Schools were spared; however, numerous other institutions were not so lucky. Brown decided to take this as an opportunity to purchase more schools. He turned his sights to California, a state Brown often visited. There he purchased fourteen apartment houses, three hotels and a few radio stations. One major purchase

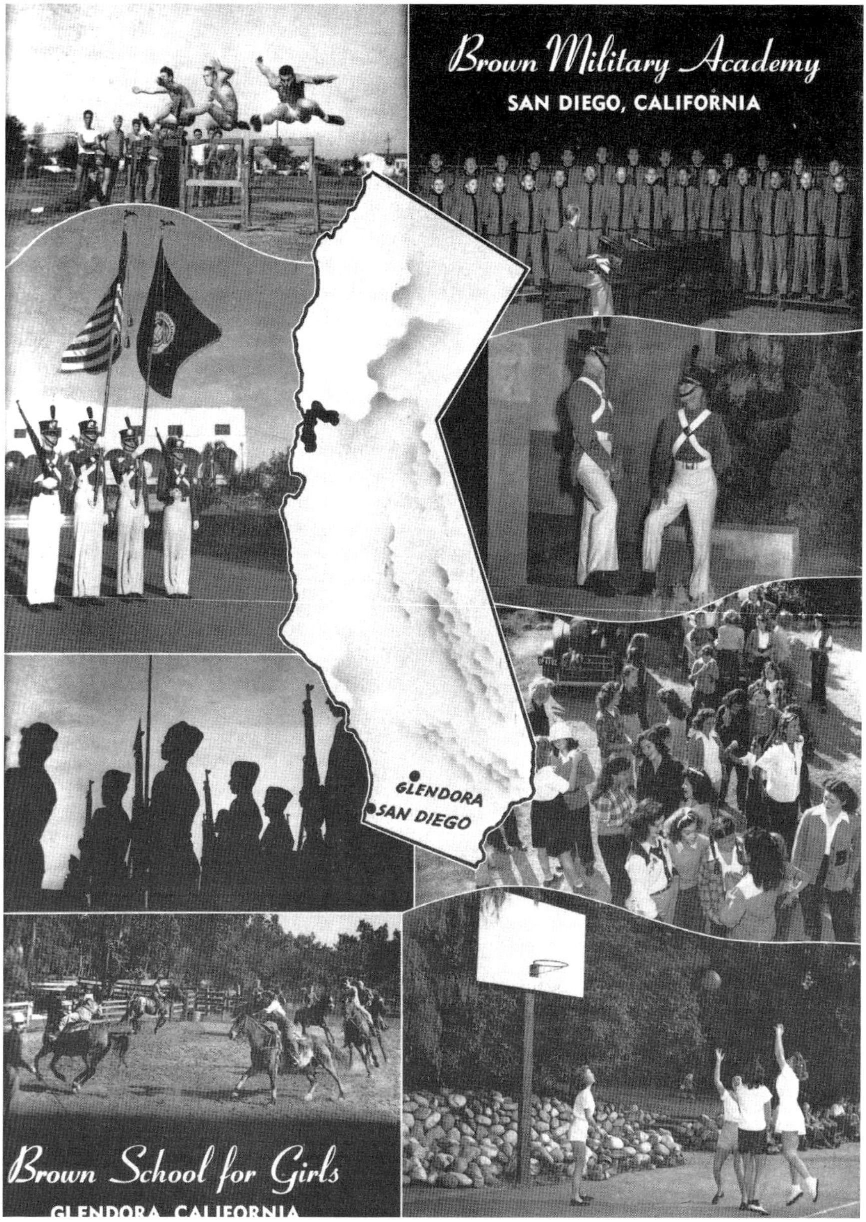

Advertisement for the Brown Military Academy in Pacific Beach and Brown School for Girls in Glendora. *Author's collection.*

was the Girl's Collegiate Institute in Glendora, which was renamed the Brown School for Girls.

In Pacific Beach, the Trust Bank, which now owned the San Diego Army and Navy Academy, was forced to manage the school in the absence of Colonel Davis. When John Brown learned of the reputation behind the Academy, he quickly snatched up the property from the Trust Bank, which was more than happy to get the school off its hands. In March 1937, Brown purchased not just SDANA but also numerous surrounding buildings and property in Pacific Beach for $125,000.

THE BROWN MILITARY ACADEMY

In 1937, the San Diego Army and Navy Academy was renamed Brown Military Academy. Len Cordoza, Class of 1941, recalled initially enrolling for SDANA in 1937 only to arrive in September to see everything was renamed Brown Military Academy. The Brown takeover was somewhat messy; the new owners weren't sure what staff and structures of the previous institution to carry over and what needed to be replaced. The identity of the new campus was somewhat schizophrenic. Cordoza commented, "I couldn't differentiate between what was San Diego Army and Navy Academy and Brown Military Academy....[S]ome of the traditions existed. They were carry overs from the San Diego Army and Navy Academy."

John Brown served as the first president of the school, but this was mostly a formality that he bestowed on himself; day-to-day operations of the school were managed by others. In March, Brown sent Ed Magourik and Charles "Mac" Griffith to take control of the school. Magourik served as the first superintendent; however, he resigned after two years following an automobile accident in which he suffered serious injuries. Barey Smiley Sr. became the next superintendent and stayed on for three years. He later became a prominent San Diego businessman. Griffith, John Brown's son-in-law, became the superintendent in 1950. Griffith purchased E.P. Wilson Company, a printing firm in downtown San Diego, and operated it from 1950 to 1977. Griffith's wife, Mary Griffith, became the next superintendent. Like Virginia Atkinson, who was vice-president of the Army and Navy Academy, this novel promotion is believed to have made Mary Griffith the only female military academy superintendent at that time.

Under John Brown, the school spent a few thousand dollars on landscaping, building projects and reconstruction. To promote his new school, Brown told the public about the merits of education at BMA and noted that it was the duty of every Christian parent to send their son to this kind of Christian environment. Brown traveled to San Diego at least once a month, where he conducted his evangelical sermons. His motto for the new school was "Training Youth to Live—Head, Heart, and Hand." Students had to live a clean life. Brown spoke against booze and "coffin nails" (liquor and cigarettes) and preached the scripture. There were chapel services every day in which cadets would sing Brown's favorite hymn, "Old Rugged Cross."

The new Brown Military Academy clung to the laurels and success of the former Academy, retaining all academic accreditation and military honors that its predecessor earned. While remaining staff helped keep the campus running along the same lines, it became evident that John Brown's teaching method and views were radically different than those of Colonel Davis. Davis was very progressive and created the school to welcome

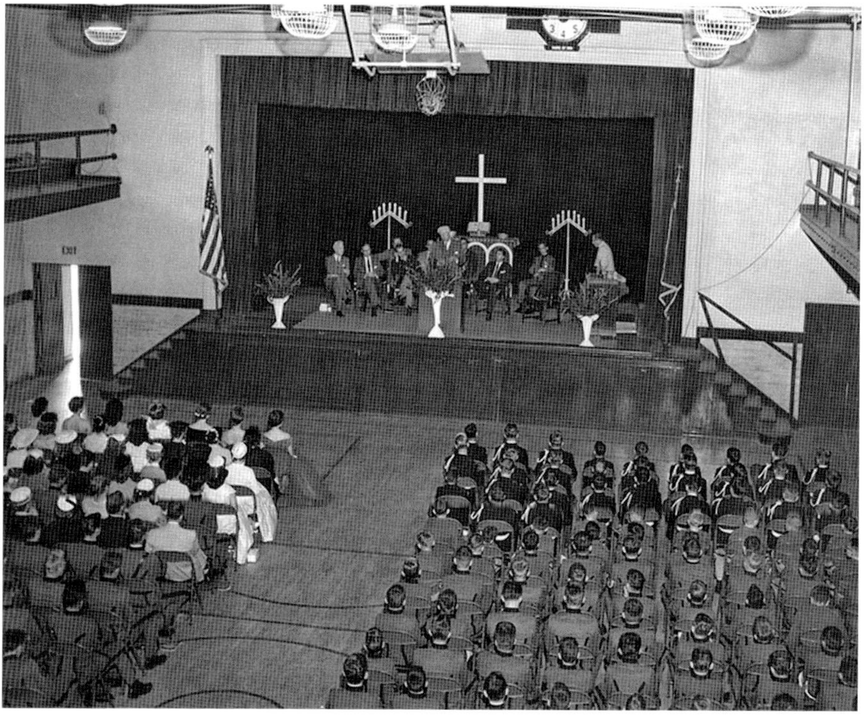

Chapel sermon given by John Brown at the Brown Military Academy. *Courtesy of Brown Military Academy Alumni Association.*

young men of any background to receive equal education in leadership and character building. Within the first two years of opening SDANA, the campus welcomed young men of countless religions, races, cultures and nationalities. This was seen as a major step forward in contrast to most schools of that era, which tended to segregate students.

John Brown, on the other hand, formed his schools as a means to promote his evangelical message. Education took a back seat and was only emphasized if it supported his beliefs. In contrast to Davis's nondiscriminatory practices, Brown founded his first school based on his segregationist views. Brown was an admirer of Booker T. Washington's Tuskegee Institute. Ironically, much like how Tuskegee Institute was exclusive to only black men, Brown wanted to create a school specifically for white students. When he founded John Brown University, he said it would be "a school that will house thousands of these wonderful descendants of pure Anglo-Saxon stock." However, while most schools in the John Brown School system followed this practice, Brown Military Academy was a considerable distance from the headquarters in Arkansas, so it saw little influence from John Brown in this regard.

In December 1938, feeling homesick, Colonel Davis retired from his position of president at the Academy in Carlsbad and returned to Pacific Beach, seeking a job at his old campus. He was welcomed back by the remaining co-workers who once worked under him. Davis's first job at BMA was as the assistant Admissions Director. To some, this was seen as the perfect position for the school's founder. As Louis Bitterlin, the last Headmaster of the Pacific Beach campus later said in his book, *The Four Academies*:

> [Colonel Davis] *was at his best when he was promoting and selling his school…the Colonel was his own best Admissions Director. He charmed the parents, particularly the mothers, while focusing his attention on the prospective cadet. His manners were impeccable, kissing the ladies' hand while extolling the virtues of military training, always with a twinkle in his eye.*

However, many saw the position of assistant Admissions Director as a slap in the face to the founder and former president of the school. Perhaps realizing that it would be better on the public relations front, Colonel Davis was promoted to vice-president of the BMA in 1939. Since the president, John Brown, only visited the campus once a month, Davis was essentially in charge of his old school again. The Junior College, which was a continuation of the San Diego Army and Navy Junior College Division, flourished, and

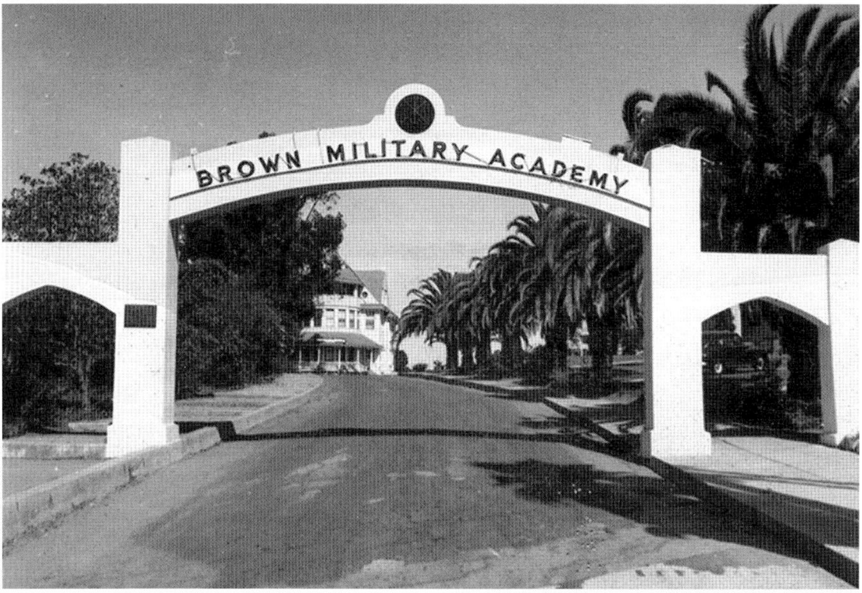

The former San Diego Army and Navy Academy Senior Archway, plastered over with the name "Brown Military Academy." *Courtesy of Brown Military Academy Alumni Association.*

a preparatory division to prepare cadets for West Point and Annapolis was added. Around this time, catalogues for the school began promoting Colonel Davis's image and his history in creating the campus, a marketing ploy to show that BMA was truly the heir of SDANA's legacy and thus worthy of all its predecessor's accreditations and honors. While Davis never publically stated his approval of this claim, as vice-president he did such a good job running the school that he caught John Brown's attention once more. In 1940, Brown stepped down and made Davis the president of BMA.

Colonel Davis also helped to bring back old traditions to liven up the cadets. He was shocked with the strict religious changes that had befallen his new campus, with the cadets' well-being and education taking a backseat to the religious proselytizing. Davis chose Pacific Beach specifically because he knew that fresh air and a social life outside the classroom were important factors to the mental development of young men. During the period Davis had been absent, Brown got rid of such social activities, even suspending dances. Alumnus Len Cordoza recounted, "Things that I found a little bit restrictive for a high school level.…[T]here were no co-ed activities, no girls on campus.…There were no social activities until Colonel Davis returned and everything just changed around. It was my Junior year. We had dances,

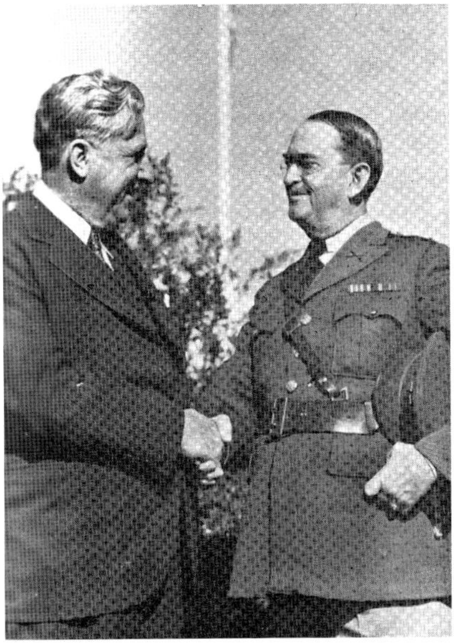

President John Brown, shaking hands with Colonel Davis, vice-president of the John Brown Schools. *Author's collection.*

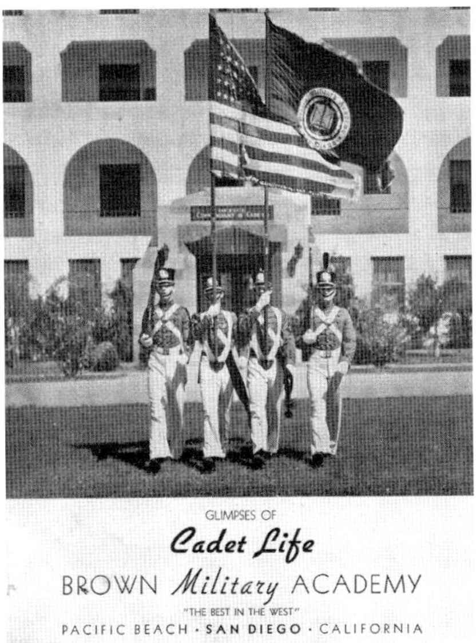

Catalogue for the Brown Military Academy. *Author's collection.*

we had more social activities. It made it a little more pleasant." For alumni like Cordoza, such social events greatly helped raise morale in the Corps.

John Brown was so impressed with the colonel's work at Brown Military Academy that in 1941 he promoted Colonel Davis to vice-president in charge of *all* John Brown Schools in the nation. This was an honor that Davis had long been trying to achieve—the operator of a nationwide school system.

During this period, when Colonel Davis served as vice-president of the national enterprise, the John Brown schools began shifting away from fundamentalism. John Brown University began welcoming its first non-white students during this era. Colonel Davis would continue to successfully manage the John Brown schools until his retirement in 1953.

In World War II, one hundred former cadets from BMA served in the military. John Brown was noted for his support of the "peace at any price" movement during this period. He said, "In the midst of this national upheaval, I stand just where I have always stood, and pledge our schools the same ideals to which they were pledged twenty-one years ago, and that is the highest standards of Christian training."

In 1944, the school set its highest record for the government's Annual Federal Inspection, with a rating of

90.5 percent. Many schools during this period rebranded themselves as military schools for wider support; however, only a few schools like Brown Military Academy and Army and Navy Academy were recognized by the War Department and received federal support.

With the success of Brown Military Academy, John Brown began to open more military schools, such as the Brown Military Academy of the Ozarks, and purchased other existing military schools, like the Southern California Military Academy of Long Beach in 1947. When John Brown retired, his son, John E. Brown Jr., became the next president of John Brown University. Like his father, he was the youngest college president at the age of twenty-six. In 1951, the Brown Military Academy of the Ozarks closed, and the boys were sent to Pacific Beach, while the girls were sent to Glendora.

THE ARMY NAVY–BROWN RIVALRY

In the history of military high school rivalries, perhaps none was more heated than the one between the Army and Navy Academy and Brown Military Academy. From athletic games to military drill, the intense competitive opposition between both academies mirrored the famed West Point–Annapolis rivalry.

Both teams were named the Warriors, as both saw themselves as the rightful heirs of the San Diego Army and Navy Academy Warriors and Pacific Beach campus. ANA's grudge came from the idea that BMA had "stolen" her founding president and property. BMA wanted to prove that it was the true continuation of Davis's legacy. However, when the much smaller ANA began to best BMA in numerous competitions, the larger school's competitive nature intensified. As one alumnus wrote, every time the BMA bus drove pass ANA's campus, the Brown cadets would sing the half-remembered lyrics in the tune of Cornell University's alma mater:

> *Far beneath the swirling waters, 'neath the muck and slime.*
> *There stands Army Navy…*

Although all competitions between the two schools were heated, it was the ANA-BMA football game where the victor would claim the biggest bragging rights. The rivalry was well known in the San Diego community and had been spoken about in legendary terms by those high school football fans and

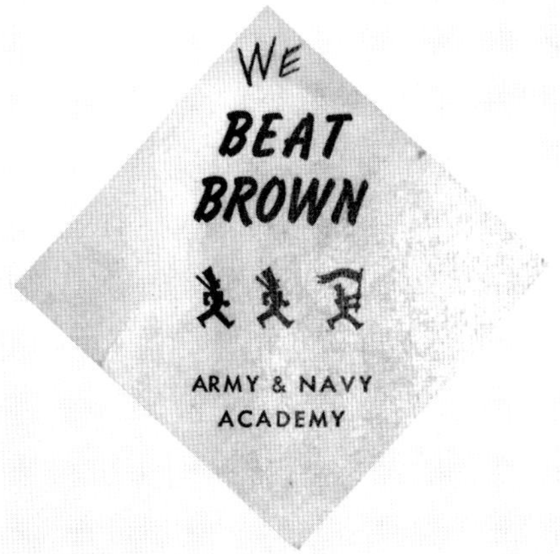

A memento from the 1956 ANA-BMA game. A cadet added "We" to "Beat Brown." *Author's collection.*

historians who remembered those highly anticipated games. The 1941 game had to be postponed so cadets from both schools could go home for Thanksgiving, but when it was held a few days later, it resulted in a 6–6 tie. Brown would go on to defeat another Warrior team, the Fallbrook Warriors, in the Southern League CIF title that season.

John "Coach" Maffucci received his degrees from UC Davis and Elkins College and served as an officer in the United States Army Air Corps. In the air corps, he also played on its basketball team. After leaving the air corps, he decided to take a job coaching the football team at the Army and Navy Academy. In 1956, when he was hired, the first thing Colonel Atkinson requested of him was that he beat Brown Military Academy that year. Colonel Atty stated that he did not care if ANA beat the civilian schools—defeating the military schools was his primary concern, and all of the military schools defeating Brown was the top priority. Maffucci promised to honor Atkinson's request and focused on strengthening the Warrior football team.

The 1956 ANA-BMA football game was highly anticipated. Bragging rights for both schools were at stake. When the big day finally came, busloads of Army and Navy Academy cadets traveled south to their school's birthplace. Nearly all the staff and faculty drove to Pacific Beach for the game as well. It is unknown whether old Colonel Davis was present for the game; however, he probably would have been supportive of both of his academies. Both sides cheered on their Warriors, which probably made things all the more confusing. This was Maffucci's first season coaching, and thanks to his training, the ANA Warriors beat Brown, 6–0. Back in Carlsbad, everyone celebrated. Colonel Atkinson was so pleased that he canceled classes and allowed cadets and staff the day off. Maffucci was promoted to the position of athletic director the following year, serving in that role for nearly sixty years. Since that 1956 victory, the tide would turn for the Army and Navy Academy, which would continue to dominate Brown in all future sports competitions.

Taps for Pacific Beach

In 1948, John Brown sold the western portion of the Pacific Beach campus, a total of five acres, to Safeway Stores for $50,000. Soon, other developers began approaching John Brown with interest in buying the remaining property, but Brown refused to sell any more land, no matter the price. On February 12, 1954, John E. Brown died in Leucadia, California. During this time, the Brown School flagship, John Brown University, had been facing serious financial obligations when retiring debentures began coming due. Brown Military Academy was actually doing quite well and bringing in profits; it was all the other Brown Schools that had become a financial burden. The Brown Military Academy of the Ozarks was closed and its students sent to Pacific Beach. With the real estate prices in Pacific Beach at an all-time high and seeing the profits of the sale as a way to financially aid the university, the Board of the Brown Schools finally sold the Pacific Beach campus for $1 million. This came as a big surprise to everyone, as enrollment had been steadily growing, with a corps numbering 603 cadets at the time of the sale. Many people believed that if John Brown had still been around, the school would have never been sold.

On June 6, 1958, Brown Military Academy closed its doors. Colonel Thomas A. Davis was invited by Superintendent Louis Bitterlin to be the honorary reviewing officer of the last parade to be held at the Pacific Beach campus on June 10. The last Battalion Commander, Melvin Stein, who later became a successful San Diego attorney, led the last dress parade. Davis was too fragile to troop the line or stand for the final pass and review. "Davis tried valiantly to stand to take the review," noted the *Evening Tribune*. "[Gen. John]

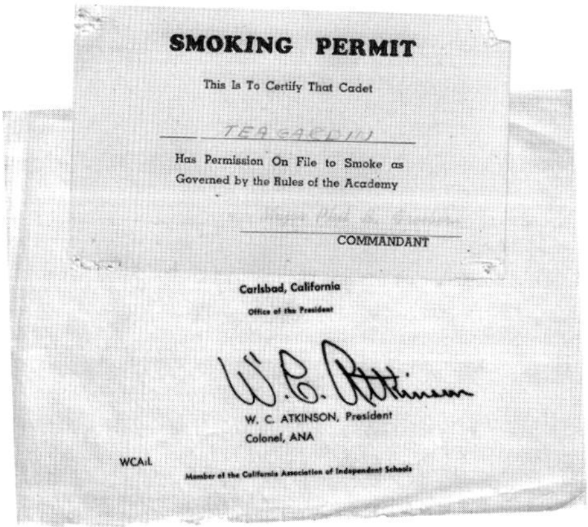

SMOKING PERMIT

This Is To Certify That Cadet

TEA GARDIN

Has Permission On File to Smoke as Governed by the Rules of the Academy

COMMANDANT

Carlsbad, California

Office of the President

W. C. ATKINSON, President
Colonel, ANA

WCA:L

Member of the California Association of Independent Schools

A cadet smoking pass from the 1950s and a letter from Colonel Atkinson. During this period, cadets were allowed to smoke on campus with permission from their parents. *Courtesy of Atkinson Museum.*

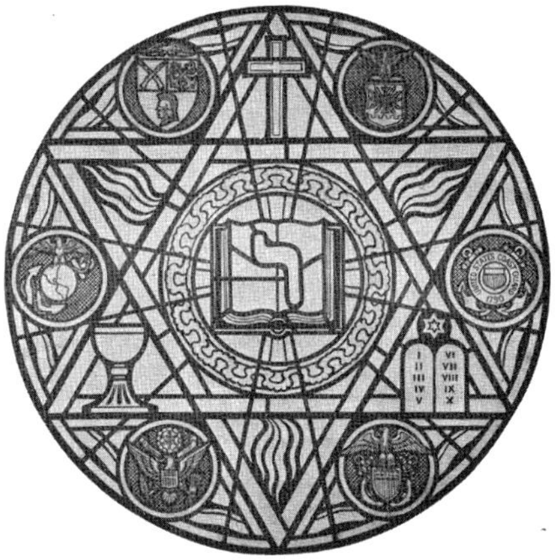

Illustration of the Virginia Powell Atkinson Memorial Chapel stained-glass window. *Author's collection.*

Harmony gripped his right elbow to steady him, but he couldn't make it." No one would know what was going through Davis's mind by this point. A half century ago, he arrived at the old Hotel Balboa to establish the San Diego Army and Navy Academy, and now he was witnessing the closing of his life's work. The colonel saluted as he watched the Cadet Corps march off the field one last time. The original Academy grounds were torn down, with no effort made to preserve the history of the location. Concrete was laid over the site, and Pacific Plaza was built.

In 2009, the alumni of Brown Military Academy and the cadets of the Army and Navy Academy held a memorial service in the original location of the Pacific Beach campus. The Brown alumni bowed their heads as ANA's color guard and bandmaster played taps, paying tribute to the ancestral home and shared heritage of both military schools.

With Pacific Beach closed, only one of all the schools Colonel Davis had established remained. Back in Carlsbad, the Army and Navy Academy was preparing for the celebration of her fiftieth anniversary. The Academy had been rapidly growing since the incorporation of Carlsbad. Despite the larger campuses and numerous advantages of nearby military schools, the Army and Navy Academy focused more on individual cadet education as opposed to increasing enrollment. This made the Academy the more appealing choice for prospective cadets. Larger military schools saw shrinking enrollment, while the small Carlsbad campus had a long waiting list.

Colonel Atkinson launched a new building project in 1953. The first goal he had in mind was to construct the new chapel in memory of his wife. Virginia Atkinson's dream was to see a chapel standing on the Academy's beach overlooking the ocean. It opened in 1958 as the Virginia Powell Atkinson Memorial Chapel.

In 1952 and 1953, the cadet yearbook won the Columbia Scholastic Press Association for two years in a row. In 1955, a new dorm was constructed and named in honor of S.J. McClendon, a doctor who had saved the lives of two cadets. In 1956, the Samuel Warfield Peterson Library was completed, named in honor of Major Pete, who loved books.

In 1957, the Academy witnessed a fire that destroyed the typing classroom. That year, a new dorm was constructed and named in honor of Vroman Dorman. In 1958, another dorm was completed and named in honor of Roy W. Hoover, a retired banker who had served twelve years on the Board. November 7 of that year saw the formation of the Mother's Club, the forerunner of the current Army and Navy Academy Patrons Association. In 1959, new classroom buildings were completed, ready for the Academy's Golden Anniversary.

Chapter 13
The Golden Anniversary

Very few institutions survive long enough to celebrate their fiftieth anniversary. Even fewer military schools achieve such a feat. However, in the last half century of her existence, the Army and Navy Academy witnessed two world wars and the Great Depression and survived moves, campus closures, bankruptcy and the retirement of her founder. Yet in those fifty years, the Academy accomplished more and earned many academic and military honors, eclipsing the countless military schools that popped up in her wake. While not the largest or having the most recognizable name, the Academy was able to stand the test of time by balancing the traditions that made her strong and adjusting to each new era. Rapid expansion is what initially doomed the Academy back at Pacific Beach and then Brown Military Academy two decades later. By limiting enrollment and committing more attention to individual cadets, the quality of education stood out against the larger military schools with public school–size enrollments. As Robert Brookhart wrote in the 1960 fiftieth anniversary edition of *The Warrior*:

> *The military system at the Academy is essentially democratic, and does not border on Militarism, but the administration and faculty believe that through military training and its attendant discipline, young men will secure the best physical development and training, and that the habits of neatness, orderliness, punctuality, and obedience will be instilled, and that there will be inculcated a deeper respect for the authority of the home, the state, and the nation.*

In an interview in the *Blade-Tribune* on November 7, 1960, covering the Academy's fiftieth anniversary, Colonel Atkinson spoke on the occasion: "Fifty years represents a long span of time and these years were filled with joy, sorrow, trials, tribulations, failure and success."

As we close the door on the story of Pacific Beach, we open a new door and celebrate the achievements of the Army and Navy Academy.

CELEBRATION DAY

In honor of the fiftieth anniversary, new white West Point uniforms were worn by seniors. This was the debut of a new tradition in which the Academy's first classmen would have the honor of wearing senior whites. Along with the dedication of the Thomas A. Davis Administration Hall, the new science buildings were dedicated in honor of patrons Mr. and Mrs. Oscar Fitzgerald Jr., who had donated the funds for the science labs. A newly built dorm was dedicated in honor of a longtime friend of the school, M.H. "Myke" Lewis. Lewis moved to Carlsbad in 1938 after retiring from his district manager job for Paramount Pictures. His wife served as Colonel Atkinson's personal secretary for more than a decade. After Virginia Atkinson passed away, she took on her role as a member of the Board of Directors. Another part of the building projects coinciding with the anniversary was the addition of the west wing to Fegan Hall, which increased the size of the mess hall and added even more rooms for boarding cadets to the upper floor.

When the anniversary weekend rolled around, the much-anticipated celebrations became the talk of the city. Carlsbad experienced a boom in visitors as old alumni, patrons and prominent civilian and military figures flooded the town for the event. Homecoming was also held that weekend, at Oceanside High School. The son of Colonel Davis, Captain Charles A. Davis, attended the ceremony on behalf of his father. A time capsule containing a recording of Colonel Atkinson and letters from the student body president and various cadets was hidden in one of the trophy cases in Davis Hall, to be opened fifty years later for the 100th anniversary.

The Warrior included a large series of articles that year dedicated to celebrating the history of the Academy. These included articles by various longtime staff members, including Captain Harold Moses. Moses came to the Academy after retiring from the army and completing his teaching education in 1946. He noted in his article for *The Warrior*, "I set out to search

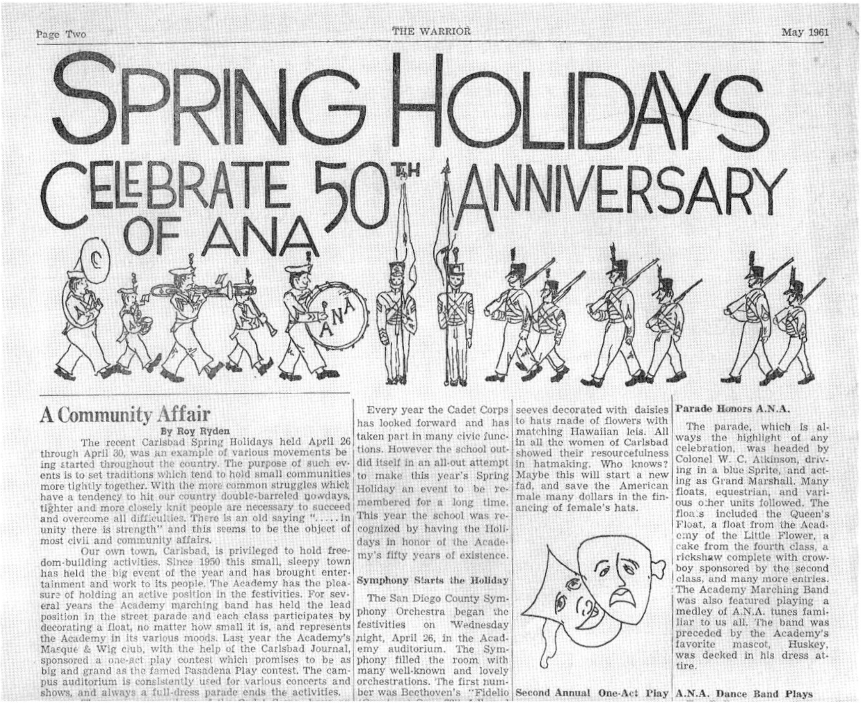

Page Two THE WARRIOR May 1961

SPRING HOLIDAYS CELEBRATE 50TH ANNIVERSARY OF ANA

A Community Affair

By Roy Ryden

The recent Carlsbad Spring Holidays held April 26 through April 30, was an example of various movements being started throughout the country. The purpose of such events is to set traditions which tend to hold small communities more tightly together. With the more common struggles which have a tendency to hit our country double-barreled nowdays, tighter and more closely knit people are necessary to succeed and overcome all difficulties. There is an old saying ". . . . . in unity there is strength" and this seems to be the object of most civil and community affairs.

Our own town, Carlsbad, is privileged to hold freedom-building activities. Since 1950 this small, sleepy town has held the big event of the year and has brought entertainment and work to its people. The Academy has the pleasure of holding an active position in the festivities. For several years the Academy marching band has held the lead position in the street parade and each class participates by decorating a float, no matter how small it is, and represents the Academy in its various moods. Last year the Academy's Masque & Wig club, with the help of the Carlsbad Journal, sponsored a one-act play contest which promises to be as big and grand as the famed Pasadena Play contest. The campus auditorium is consistently used for various concerts and shows, and always a full-dress parade ends the activities.

Every year the Cadet Corps has looked forward and has taken part in many civic functions. However the school outdid itself in an all-out attempt to make this year's Spring Holiday an event to be remembered for a long time. This year the school was recognized by having the Holidays in honor of the Academy's fifty years of existence.

Symphony Starts the Holiday

The San Diego County Symphony Orchestra began the festivities on Wednesday night, April 26, in the Academy auditorium. The Symphony filled the room with many well-known and lovely orchestrations. The first number was Beethoven's "Fidelio

seeves decorated with daisies to hats made of flowers with matching Hawaiian leis. All in all the women of Carlsbad showed their resourcefulness in hatmaking. Who knows? Maybe this will start a new fad, and save the American male many dollars in the financing of female's hats.

Second Annual One-Act Play

Parade Honors A.N.A.

The parade, which is always the highlight of any celebration, was headed by Colonel W. C. Atkinson, driving in a blue Sprite, and acting as Grand Marshall. Many floats, equestrian, and various other units followed. The floats included the Queen's Float, a float from the Academy of the Little Flower, a cake from the fourth class, a rickshaw complete with crowboy sponsored by the second class, and many more entries. The Academy Marching Band was also featured playing a medley of A.N.A. tunes familiar to us all. The band was preceded by the Academy's favorite mascot, Huskey, was decked in his dress attire.

A.N.A. Dance Band Plays

The final special edition issue of *The Warrior*, which was printed on yellow paper for the 1960–61 school year to celebrate the Academy's fiftieth anniversary. *Courtesy of Atkinson Museum.*

for a good private school to which to apply my trade. The University of California had told me of the Army and Navy Academy and of its growing reputation so I set my sights for this school."

Moses is best known for his work as director of the Army and Navy Junior Division. Along with his work for the Junior School, he was known for his work with the Academy's tennis team and yearbook. He oversaw the cadet production of the yearbook, which had just been awarded first place by the Columbia National Press Association two years in a row.

The celebration continued well into 1961, when the Carlsbad Spring Holiday Parade made Colonel Atkinson the grand marshal of the event in honor of the anniversary. The community also received a surprise that year when it was announced that the San Diego Chargers were considering ANA as a possible summer training site. At the dawn of the next half century, in an address printed in *The Warrior*, Colonel Atkinson said:

> *Fifty years is a long span of years in the life of an individual or an institution. Both are alike, however, as related to the ultimate end of the span in so far as qualities of success, dignity, and ideals are concerned.... Twice in my service with the Academy have I seen it reap success and behind these two periods of success stood always a demanding dignity and challenging ideals. Another fifty years will pass—and still another, and on through the ages Army and Navy Academy will stand as a successful institution for I know that its ideals will ever be perpetuated.*

FAREWELL TO THE FOUNDER

Now older and somewhat frail, Colonel Davis, who couldn't even stand during the last parade review for the closing of Brown Military Academy two years before, was invited by Colonel Atkinson to attend the festivities in Carlsbad to honor the anniversary of his most successful school.

By this time, it had been more than twenty years since Davis left his presidency in Carlsbad to pursue a job back at Pacific Beach. This was no longer Davis's school. The Davis Era had long passed; this was the height of the Atkinson Era. Atkinson had transformed the Army and Navy Academy into something new while keeping the spirit of the original principles established by Davis; the current school was now something unrecognizable to Davis. The Academy had seen massive development around the old former hotel. Large buildings loomed over the vast seaside campus, and in the center, a massive gymnasium/administration hall served as the new face of the Academy, a landmark in Carlsbad that welcomed visiting prospective students and parents alike.

It is not known what the relationship between the two colonels was like. While it is a personal observation, in all my research and reading of historical documents, I never once read a mention of Davis and Atkinson together. Perhaps this was deliberate. When compiling history for the fiftieth anniversary, Atkinson noted specifically that it was the younger John Davis whom he first met and who convinced him to take the bandmaster job in Pacific Beach. Alumni, who remember Atkinson when talking about history, recall the colonel only briefly mentioning Davis and the founding and then skip over his era and Pacific Beach, noting that this was all in the past and that people needed to look forwards.

In his book, *The Four Academies*, Louis Bitterlin also mirrored my observation:

> *Although they were not the best of friends, Colonel Davis and Colonel Atkinson had much in common: They epitomized what a Southern gentleman should be. They were super patriots, and they wanted very sincerely to turn out cadets who, you might say, were Christian Warriors prepared to lead the way.*

Regardless of prior relationships, towards the end of his life Davis and Atkinson became cordial pen pals. As the once massive Brown School System began to dissolve, Colonel Davis was at least pleased that he had left his mark with the recent success of his school in Carlsbad. Davis still lived in Pacific Beach, near the old campus, in the same old house at 4637 Lamont Street he first moved into when he came to California back in 1910. Having been retired from his position at Brown since 1953, Davis intended to live out his remaining days with his wife in that very same house. The campus of his old school surrounded the property. From his window, Davis would be forced to witness the buildings of his old school being destroyed and built over as if they never existed. Around his house emerged a parking lot and strip malls. All his old friends from Pacific Beach moved over to the new campus in Carlsbad one by one. Nothing is permanent.

As November neared, Atkinson was intent on having the founder present for the celebration of his Academy. He personally wrote to Davis on numerous occasions with such a request. Eventually, Davis responded back in a letter dated November 9, 1960:

> *I take this opportunity to congratulate you again on the success of the Academy under your direction. You are headed for greater success….My prayer is that God's richest blessings will rest upon all who have ever been connected in any way with the old Academy to which I devoted so many years of my life.*

Davis had not yet made a decision to make the trip to Carlsbad to visit his old school and old friends. Four days later, he sent another letter:

> *In my weakened condition and my inability to walk around in the room without some kind of support, I really feel that it would be unwise to make the attempt, despite my truly yearning desire to be there and to see my old friends. I assure you that it is a keen disappointment.*
>
> *I request again that you extend my greetings to the old boys who may be in attendance and let them know that I should be happy indeed to see them if they could call at my home before returning to their own home.*

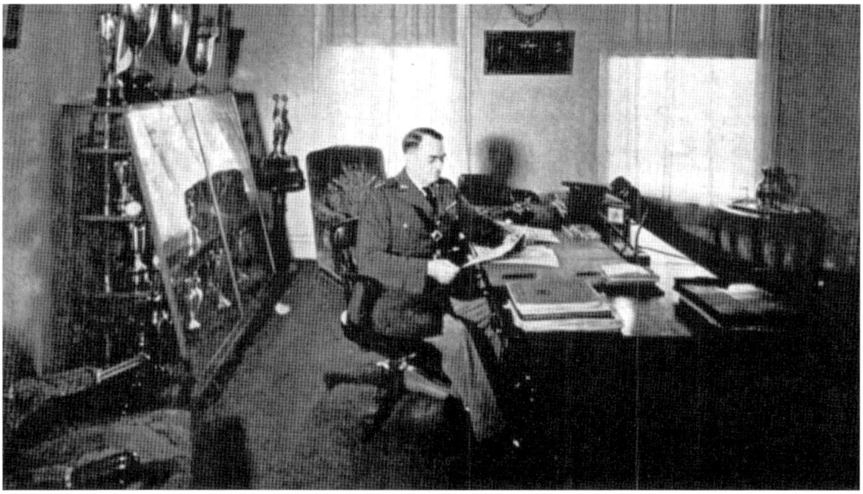

Colonel Davis in 1936, during the last year the San Diego Army and Navy Academy operated in Pacific Beach. *Courtesy of Atkinson Museum.*

Davis is not only remembered as the Academy's founder but also as the best Admissions Director, whether it was with El Paso, Pasadena, Carlsbad or Brown. So it was only fitting that during the anniversary celebrations, the unnamed admissions building, the face of the Army and Navy Academy, was dedicated as the Thomas A. Davis Administration Hall.

Three years later, on February 12, 1964, Colonel Thomas Alderson Davis passed away peacefully at the age of ninety in the same house he lived in for the past fifty-four years. A service was held for him at Green Mortuary on February 18. His family requested no flowers but suggested that contributions be made to the San Diego Children's Hospital in his honor. Davis was buried in the Old Cathedral Mausoleum at Greenwood Memorial Park. His wife was later buried beside him.

The 1936 dedication in the last yearbook to be published from Colonel Davis's pride and joy, the San Diego Army and Navy Academy at Pacific Beach, summed up the old founder best:

> *Dedicating one's life to the understanding of a young lad's heart is no easy task, but this is child's play as compared to guiding the destiny of a surging group of adolescent, military-minded young men.*
>
> *It might well be said that the state and the institution are exceedingly fortunate in having a man like Colonel Davis, gentle and tolerant, yet*

fearless and original in thought, to organize and develop an educational foundation as symbolized by this Army and Navy Academy.

Under his unfailing guidance and continual search for new ideas, the Academy has become the foremost of its type on the Coast. The ability to become a personal adviser on the most trivial provocation, has endeared him to all with whom he has come in contact.

Last of the Red Apples

During that same year, Joe Orozco started working in the mess hall. In an interview with him, he said that he could recall a cream meat dish called "mung" that they served to the cadets. "One thing I could fondly remember were the red apples that remained from the Red Apple Inn." The red apples dotted every piece of furniture in the old inn—the gates had apples made of clay, apples were forged into the iron electric fixtures and even all the china plates had the red apples fired on them. Back when the Academy relocated to the new campus, there was no money to replace the china, so they were put into use, hoping that they would "vanish through attrition." Every time a plate fell, the crashing sound ringing throughout the mess hall, the cheer could be heard by everyone. "Every time Col. Atkinson heard a plate break he would say thank goodness," noted Joe.

Years before, while working in the mess hall, Joe repainted red apples on the light fixtures that can still be seen today. Joe recalled his memory of those apples, how Mrs. Atkinson would warn him about painting them. She said, "Joe, Col. Atkinson doesn't like those apples he is going to get angry." But Joe continued painting the apples, and the first thing Colonel Atkinson said when he walked by was, "Who painted those apples?" Mrs. Atkinson noted that it was Joe, but the colonel replied, "That's okay. Joe is my favorite little guy here."

Through the years, more and more plates broke until there was apparently only one left. The Academy celebrated as the last piece of red apple was ceremoniously broken. Later, Joe found another box of red apple plates in the back of the mess hall and put them to use. This annoyed Colonel Atty, who would continue to cheer whenever they broke. The plates were made from china and would crack in the dishwasher because they were so old. Finally, when there was only a handful left, the colonel told Joe to throw them out. They were never seen again, and with them went the last piece of the famous inn. Joe noted that it was a shame he didn't preserve one so a piece of the Red Apple Inn could be remembered.

CANINE MASCOTS

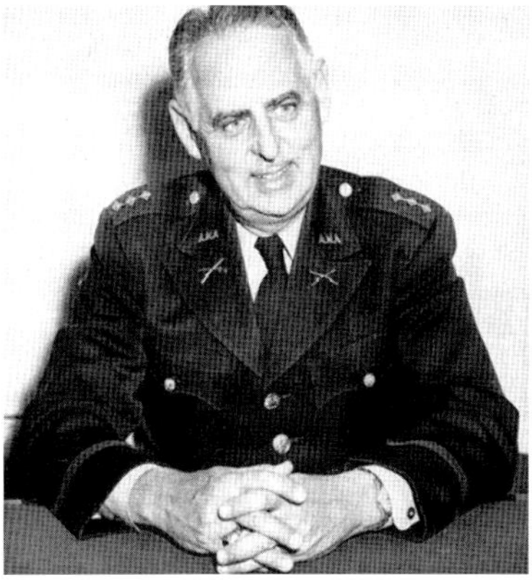

President Colonel Atkinson in the 1960s. *Courtesy of Atkinson Museum.*

Cadets in the 1960s remember the Academy having dog mascots wandering campus and trooping the line during parades. The original mascot was a young boxer named Husky, which was a gift to the school. He was seen around campus, often hanging out near Colonel Atkinson's office.

After Husky, the next mascot was an English bulldog named Sarge, gifted to the school in the mid-1960s. He lived with Colonel Atkinson and often was in attendance at sports games. He would also attend weekly parades in his own dress tunic, complete with a sergeant rank.

The last of the three mascots was a German shepherd that just showed up on campus one day. The cadets took care of him, providing him with food and a place to sleep. It was said that one day Colonel Atkinson saw a group of cadets sneaking the dog into the dorms and had the Commandant look into the matter. The Battalion Commander reported to the colonel to explain the situation. Colonel Atkinson appreciated the cadets' selfless act and allowed the dog to remain as part of the Corps. As an article in *Army Navy* magazine put it, "It was clearly a case where the Cadet Corps adopted the dog and the dog adopted the Cadet Corps." The German shepherd was named Sergeant Ace and was often seen camped outside the Commandant's Office. It is noted that he routinely patrolled the campus with the Night TAC Officers.

RETURN OF JROTC

In 1966, Colonel Atkinson signed a contract with the U.S. Army to bring JROTC back to the Academy. In November 1966, Major Donald Herbert Cline joined the Academy as the first Senior Army Instructor of the Carlsbad campus and was the first active duty officer from the army to set up a Junior Reserve Officers' Training Corps program on the new campus. The Academy had been under the National Defense Cadet Corps for decades

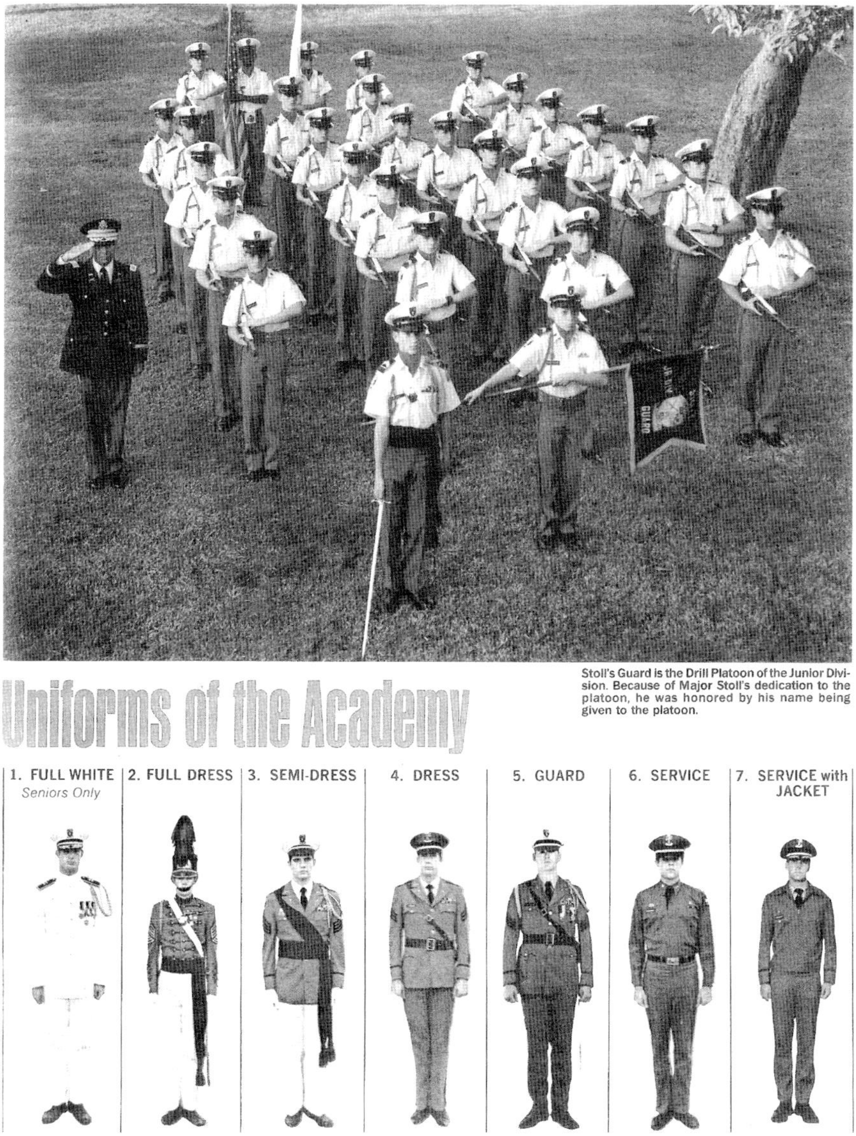
Stoll's Guard is the Drill Platoon of the Junior Division. Because of Major Stoll's dedication to the platoon, he was honored by his name being given to the platoon.

Uniforms of the Academy

1. FULL WHITE *Seniors Only*
2. FULL DRESS
3. SEMI-DRESS
4. DRESS
5. GUARD
6. SERVICE
7. SERVICE with JACKET

A page from a catalogue for the Army and Navy Academy, showing the middle school Stoll's Guard on the top and beneath, the different types of uniforms a cadet would need to wear. *Author's collection.*

since her move but was finally approved by the Department of the Army as a military institute under the JROTC program. Major Cline arrived at the Academy after his first tour in Vietnam. He stated, "I thought I had arrived in Heaven." He was thrilled by the fact that he was able to work with young men in a military atmosphere since he was a graduate of West Point. Major Cline received his Congressional appointment to West Point in 1953 and graduated as a member of the Class of 1957.

Major Cline said that he would never forget the warm welcome he received the day he reported to Colonel Atkinson. Colonel Atty told him, "It is an honor to have a West Point graduate as the first Military Instructor at the Academy. We pride ourselves as being the 'West Point of the West' and I am sure you will help us to be just that." Major Cline became a personal friend of Atkinson's and maintained contact with him after he left the Academy to serve his second tour in Vietnam. He described his time at ANA as "two years in paradise." Later, when Cline visited his family in Los Angeles, he looked forward to a ninety-mile drive to visit his friends at the Academy. "It was like coming home."

PART V

THEY JUST FADE AWAY

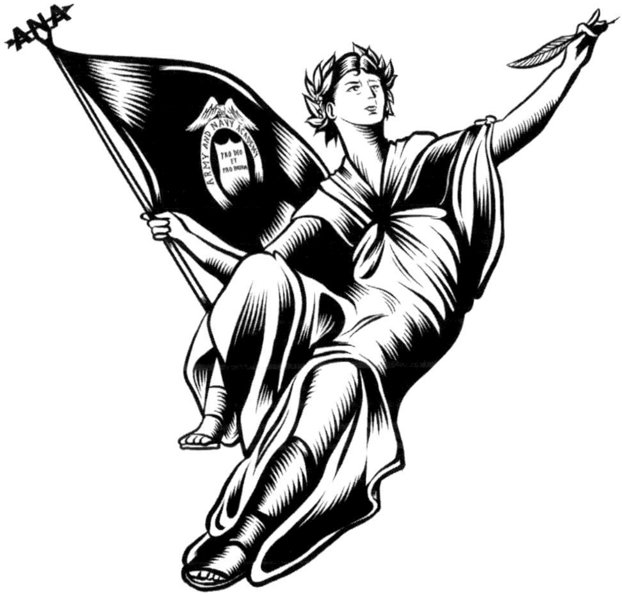

The Academy's Turmoil. Artwork by Alexander Mui.

Chapter 14

All Glory Is Fleeting

Protest erupted across the nation as public support for the Vietnam War was at an all-time low. Many of the protests were concentrated in California, the Academy's home. As the antiwar movement intensified and events like Watergate gave rise to government dissatisfaction, military schools began to close their doors one by one.

In previous eras, a major war normally meant a spike in the birth of new military schools. With the Vietnam War, this was not the case. In fact, no new military school would be created after the staff of Eastern Military Academy (1944–79) in New York split and formed the Marine Military Academy (1965) and MacArthur Military Academy (1966–77). The only other military school formed during this period to survive to the present was the Florida Air Academy (1961).

While most military schools faced their inevitable fate of closure, the Army and Navy Academy was blessed with the Andersons' gift. With the Andersons gifting all land to the Academy, there were no loans needing to be paid or land needing to be leased, which all other military schools had to contend with. However, the Academy was far from safe and endured a long struggle during this period.

California, the heart of the antiwar movement, was not the best place for a military school during this period. Enrollment dropped rapidly as protests against the war intensified. ANA was losing money year after year. The Academy's attorney, Vroman J. Dorman,, sold large portions of the campus to balance the financial loss. This land included two large athletic fields known as the Point and the Horseshoe. The Point was used as the Academy's baseball field; however, it was divided by a street that ran down the middle. When running to second base, cadets would have to watch for cars. The Point was converted into a residential area and today comprises houses, apartments and hotels. This sale was not without its own controversy. Rumors began to spread that Dorman was taking advantage of the Academy and the aging Colonel Atkinson.

SUCCESSORS TO BROWN MILITARY ACADEMY

Back at Pacific Beach, the grounds of the former SDANA/BMA were completely demolished. In the spot of the former campus now stands Pacific Plaza, a mundane collection of shops and a strip mall.

The John Brown Schools, under John E. Brown Jr., tried to remain steady throughout the Vietnam War and the antiwar movement. The flagship school, John Brown University, remained strongly Christian and politically conservative, with students joining groups like Campus Crusade for Christ as opposed to peace movements. However, the administration of the John Brown Schools, 1,500 miles away in Siloam Springs, was completely out of touch with the problems faced by the Californian schools purchased in the 1930s and 1940s. Even when John Brown was still alive, he appeared very disconnected with some of his California properties. One alumnus recalled John Brown visiting one of his military schools for Christmas dinner. Wanting to find a reason to quickly depart the event, he asked one of his associates to pinch his young boy. When the boy started crying, Brown and his associates quickly departed.

In 1937, when Brown purchased the San Diego Army and Navy Academy's Pacific Beach campus, he also purchased the Girl's Collegiate Institute in Glendora. The Glendora school was renamed Brown School for Girls. The same year, Brown also formed the Brown Military Academy of the Ozarks, in Sulphur Springs. In 1951, Brown decided to gift the Sulphur Springs campus to the Wycliffe Bible Translators and sent the boys from the school to Pacific Beach and the girls to Glendora.

The Brown School for Girls saw competition from another girl's school in the area. After realizing the high property value of the Pacific Beach campus, the Board of Brown Schools decided to discontinue the Brown School for Girls and relocate Brown Military Academy to the Glendora property.

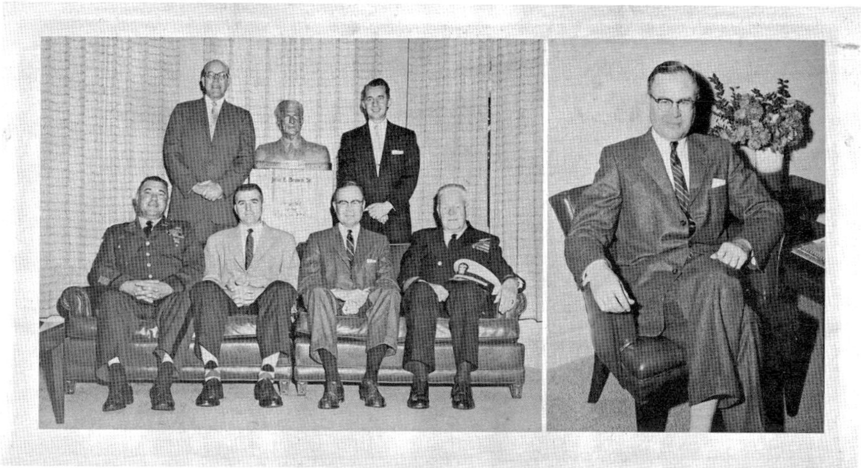

Left: The Brown Military Academy Executive Committee. Bitterlin is seated in the front row, *center right*, in front of a statue of founder John Brown. *Right*: Louis Bitterlin as the president of the San Diego Military Academy. *Author's collection*.

The new BMA in Glendora opened in 1958. The new school was relatively small, with enrollment size said to be around 125 to 150 cadets. Camp Buddy, named after the nickname John Brown gave his son, operated as a coeducational summer camp. While a few cadets from the Pacific Beach campus continued on to the Glendora incarnation, most of the staff were new. Most former Pacific Beach cadets and staff, however, would continue on to another school that many felt was the true successor.

LOUIS J. BITTERLIN

Louis Bitterlin was born in Guaymas, Mexico. He was raised and educated in Los Angeles and attended Santa Monica High School, where he excelled in tennis. Bitterlin ranked number one in singles for the 1932 CIF Southern Section Championship. He later graduated from Stanford University, where he was a member of Alpha Delta Phi and earned his master's degree in political science.

Bitterlin began a teaching career at the California Military Academy in Los Angeles, despite never having served in the military. "He was a pacifist, a conscientious object," said his daughter, Gretchen. "It was a moral decision he made after college."

Despite identifying as a pacifist, he became very fascinated with military-style education. While at the California Military Academy, he taught classes and coached football, basketball and baseball. He later taught at the Southern California Military Academy in Long Beach, another school purchased by John Brown.

In 1947, Bitterlin moved to San Diego, where he became an instructor at Brown Military Academy and then Dean of the Junior College. Bitterlin was Catholic, something unheard of in the Brown Schools, which held strongly to evangelical views. Bitterlin joked that he tried to convert the campus chaplains to Catholicism, to no avail. In 1952, Bitterlin became the headmaster of BMA. When John Brown died in 1954, Bitterlin also took on the role of superintendent. As superintendent, he decided to eliminate the Junior College to make more room for the High School, which was growing in enrollment. He felt that a Junior College and High School were not compatible on the same campus.

During his time, Brown was successful; enrollment was at an all-time high with a Corps of 603 cadets and was steadily growing. The summer camp program was very popular, and cash was consistently flowing in. When it was announced that the Pacific Beach campus would be sold and BMA would relocate to Glendora, it came as a shock to everyone. The Board of Brown Schools stated that the decision to close the school was simply a matter of economics. Bitterlin was not pleased to hear this after investing so much work into the school.

The Pacific Beach campus closed its doors on June 6, 1958. While the Board of Brown Schools advised cadets to continue on to Glendora, Bitterlin felt that there needed to be a proper continuation of BMA in the San Diego area, not one more than one hundred miles away. He contacted all the parents of boarding and day cadets and proposed to them an idea of establishing a new school in the local area. He learned that many parents were interested in enrolling their sons at his new school. Bitterlin knew that he had to hurry, as he had only one summer to create a new school in time for the next school year.

In Solana Beach, he obtained the Las Flores Inn. The inn was an ideal location for a school, exactly equal distance between Pacific Beach and Carlsbad. The inn was already equipped with a swimming pool, kitchen, dining room and guest rooms for boarding cadets. There was enough room for parade fields, athletic fields and expansion with twenty acres of land. The one problem was that the inn did not provide buildings for classrooms; however, Bitterlin had already solved the problem. He found surplus portable classrooms available in Pacific Beach and purchased them for $500 apiece. Six of the rooms were divided to make twelve classrooms, and the other two were used as a band room/science lab.

The property in Solana Beach, along with the Las Flores Inn, cost Bitterlin $200,000. While the inn was near the Del Mar Fair Grounds and Horse Racing Track, the original owners of the inn sold the property, as it only brought in profits during the summer months during the horse racing events. Bitterlin continued to operate the inn during the summer months to generate income.

Bitterlin hired William Nietfeld, one of San Diego's leading public relations figures, to promote the new academy as well as the inn. Within three weeks, on Memorial Day weekend, the inn was overflowing with guests. For four summers afterward, the inn continued receiving many guest from the racing crowd. In 1962, Bitterlin ended the practice of room and boarding when he opened the summer camp program, Camp Cabrillo. But even with the inn closed, people still arrived for years afterward seeking a room to stay for the night.

Louis Bitterlin had long been fascinated by military schools. Despite never having served in the military, the history of the military school institution, especially in San Diego, greatly interested him. "Military academies were very prominent in the San Diego area's educational scene. It's a fascinating, valuable history that goes back to 1910," noted Bitterlin on the subject.

He admired the work Colonel Atkinson had done with the Army and Navy Academy and even saw his new school as a parallel to the story of the Carlsbad military school. When Colonel Davis was forced to leave the Pacific Beach campus, he and a portion of the San Diego Army and Navy Academy staff ventured north to buy an inn to continue his school. In much the same way, when Bitterlin was forced to the leave the Pacific Beach campus, he and the same number of the Brown Military Academy staff ventured north to buy an inn to continue his school.

THE SAN DIEGO MILITARY ACADEMY

On September 21, 1958, Louis Bitterlin officially declared the San Diego Military Academy open. Present at the dedication was Colonel Atkinson, who congratulated Bitterlin and wished him luck with his new academy.

The Army and Navy Academy donated its old scoreboard as a gift to the San Diego Military Academy. Bitterlin's new school was now fourteen miles away from the Army and Navy Academy, which continued the previous rivalry with Brown Military Academy in sports and military drill competitions.

Unlike the Army and Navy Academy, San Diego Military Academy was a for-profit school owned by investors and stockholders. The people who made up the original investors, many former Brown staff, became known as the founding fathers of the new academy. Bitterlin served as chairman of the Board and Headmaster of San Diego Military Academy. His mission for the new school was "not to make soldiers" but to "offer structure and supervision in a controlled environment."

San Diego Military Academy opened with an initial enrollment of ninety-one cadets. The school had a smooth start considering the brief time spent in its establishment. This was made possible due to the fact that most of the faculty and staff had formerly worked with one another at Brown Military Academy. Also, many of the principal investors held staff positions on campus and thus had personal interest in every aspect of the school's operation.

Future expansion, on the other hand, was limited due to the gas tanks and waterlines running through the Rancho Santa Fe community.

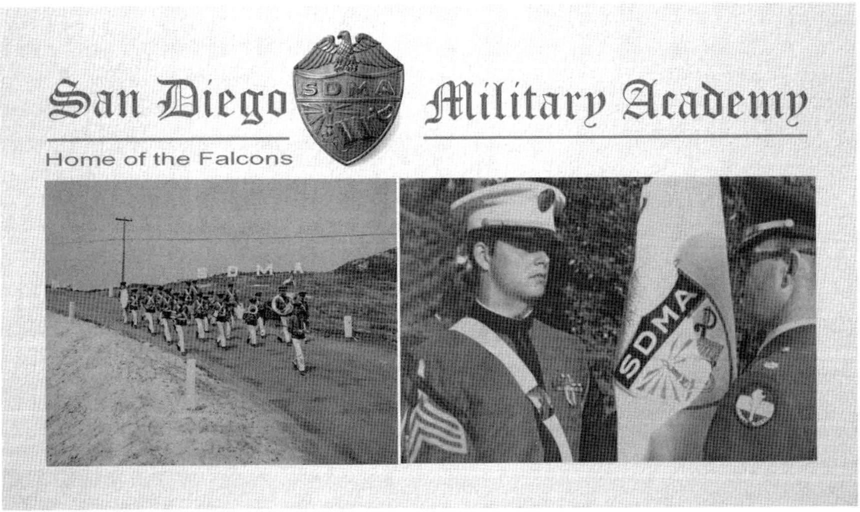

Left: The SDMA band marches down Academy Drive in 1959. A large sign with the letters "SDMA" stands in the background. Legend has it vandals would move the letters around from time to time. *Right*: Cadet with the SDMA colors. *Courtesy of SDMA Alumni Association.*

However, with a growing student population, Bitterlin initiated a building program. Architect Russell Forrester designed the buildings without fee. Bitterlin had a large two-story dormitory and classroom building built for the Lower School. This was then followed by an administration building, an infirmary, additional classrooms, a post exchange, a cadet store and a multipurpose building that housed the library, the Commandant's Office, the armory and additional classrooms. The terrain over the Santa Fe Hills was undulating, so Bitterlin had the ground leveled for the football field, parade ground, baseball diamond, the two tennis courts, the basketball court and the rifle range. All of the financing for these buildings came from the San Diego Federal Savings and Loan Association, under the leadership of Jack Thompson. In 1962, the library was completed.

In 1964, Bitterlin became president of the Del Mar Rotary Club and the Western Association of Private Schools. In 1966, the Brown Military Academy in Glendora closed, yet the San Diego Military Academy was still flourishing. However, since the San Diego Military Academy was a for-profit organization, the school was not allowed to receive tax-deductible gifts or donations. Also, this meant that it could not receive any government or state grants or programs.

In the early 1970s, the school had to pay $10,000 for liability. Since San Diego Military Academy was a boarding school, that meant there was an added responsibility of watching over the cadets twenty-four hours a day. This included completely abiding by numerous safety codes set out by the insurance company and state, involving school bus transportation, contact sports like football, preparation of meals, swimming pool safety and extra caution at the rifle range. The antiwar movement also made its presence discomforting to the local area.

When it came time to finally renovate the old buildings, the Board of Directors found it necessary to have an extensive rebuilding project. Thus, rather than invest more money, the Board decided to sell the school and cash in on the investments.

Bitterlin viewed this situation in the same manner as Brown Military Academy's closing in 1958. The sale of San Diego Military Academy was not based on low enrollment or financial stress but rather just as a matter of economics by the Board.

In 1978, the San Diego Military Academy was sold to the Christian Schools of San Diego, which bought the property for $1 million. With that came an end to the twenty-year-old school. Bitterlin remained the sole president of the San Diego Military Academy from its founding to closing.

The school grounds are now currently owned by Santa Fe Christian High School, named after the Santa Fe Hills on which the San Diego Military Academy was originally built over.

In 1994, Mr. Bitterlin published his book, *The Four Academies*, which was his love letter to the history of military education in San Diego. The book outlined the history of the four military academies of San Diego: San Diego Army and Navy Academy, Brown Military Academy, Army and Navy Academy and San Diego Military Academy. Soon after he published the book, he passed away from a stroke at the age of eighty-eight. On May 3, 2003, the remaining Brown Military Academy and San Diego Military Academy alumni gathered to dedicate a memorial plaque at the Glendora campus site to honor both academies.

THE HALL OF FLAGS

Back in Carlsbad, the Army and Navy Academy was riding out the antiwar movement brought on by the Vietnam War. Major General Hugh H. Milton, Secretary of the Army, visited the campus in 1969. The High School drill team was named Milton's Guard in his honor. In 1971, the Academy won the gold star for Honor Unit with Distinction. That year was also renowned for the Senior Class Hall of Flags project. The president of the Class of 1971 came up with the project, and it was approved by the senior class in December 1970.

From January 1 to April 15, each member of the senior class was assigned a state and wrote to the governor of that state. Each cadet requested a state flag and information of that particular state to be sent to the Army and Navy Academy. Twenty-eight governors sent their state flags to the Academy, with some being flown at their various state capitols before being sent.

Other flags were also sent. Prime Minster Trubaul of Canada sent a representative who presented the Canadian flag. Consul Monsieur Daniel LaBrosse presented the French flag. Alumnus Mikko Pimia, Class of 1967, ANA's first foreign exchange student, presented the Finnish flag. As word of the project spread, prominent figures began sending their letters of support, including President Richard Nixon, Vice President Spiro Agnew, General Omar Bradley, former president Lyndon B. Johnson, John Wayne, the head of NASA and numerous other prominent figures.

The person most supportive of the project was California governor Ronald Reagan. The governor sent numerous letters to the Army and Navy Academy showing his support. He first learned of the project when he was told that a state official had turned away the cadet's request for a flag as there was no reserve budget for such a request. Governor Reagan and the women of the California Republican Party donated the money to cover the cost of the flag. The Bear Flag was flown over the state capitol in Sacramento and sent to the Academy, along with a letter from Governor Reagan. This was a great honor, considering California is the Academy's home state.

On June 5, all the flags were presented along with the representatives of the various states. Governor Reagan sent a representative down to the Academy on his behalf. On June 21, the Hall of Flags was nominated for an award from the Freedom Foundation in Valley Forge.

End of an Era

In 1972, near the end of the Atkinson Era (1943–80), the Academy won her first CIF Football Championship; this was the same year Colonel Atkinson chose to retire. He was succeeded by his son, William C. Atkinson Jr., who served as mayor of Carlsbad from 1965 to 1968. As mayor, he oversaw the annexation of La Costa to Carlsbad. He served as vice-president of the Academy for more than twenty years and as president for two years. During his short two-year term, the school started facing financial problems, deteriorating standards and various drawbacks.

In 1974, Colonel Atkinson was convinced to come out of retirement to keep the Army and Navy Academy on the right track during this critical time. In 1980, he was able to witness the basketball team win its first CIF Championship. At the end of the year, Colonel Atty once again announced his retirement, this time permanently. He was succeeded by Colonel Gilbert N. Brookhart.

The effects of the Vietnam War were evident. Reacting to the public's negative reaction to military schools during this period, Colonel Atkinson said, "If a boy comes to us, he does things our way….If someone says we are too strict on their boy, they are free to leave."

While the Army and Navy Academy carried on with business, other military schools weren't so lucky. The nearby Black-Foxe Military Institute (1928) closed in 1968. The oldest military school in the nation, Charlotte

Hall Military Academy (1774), closed in 1976, two centuries after its founding. The two schools Colonel Atkinson had previously worked at—Greenbrier Military Academy (1812–1972) and Castle Heights Military Academy (1902–86)—also closed.

In a reverse of what occurred after World War I, when civilian boarding schools rebranded themselves as military schools, during the Vietnam War numerous military schools dropped the military aspect, becoming nonmilitary schools. Notable examples include Morgan Park Military Academy (1863) in Chicago, which ceased operating as a military school in 1958 and was renamed Morgan Park Academy, eventually dropping the boarding aspect of the school as well; Georgia Military Academy (1900), which became a coeducational boarding school in 1964 and was renamed Woodward Academy; Sacred Heart Military Academy (1955), which became the nonmilitary LeMans Academy in 1968 after relocating from Wisconsin to Indiana; Baylor School (1893), which became a military school during World War I and remained one until 1971; Harvard School for Boys (1900), which discontinued both boarding and military aspects of the school in the late 1960s and later merged with the Westlake School for Girls (1904) to become Harvard-Westlake School in 1991; Harker Academy (1893) in San Jose, which dropped the military and boarding portions of its school and became Harker Day School in 1972; and the Palo Alto Military Academy (1919), which was founded as the nonmilitary Manzanita Hall (1893) and merged with Harker Day School in 1973 to become the nonmilitary Harker School.

However, even becoming a nonmilitary school did not spare numerous schools from financial issues. Marymount Military Academy (1923) in Tacoma, Washington, ceased operating as a military school in 1971 and became a private boarding school; however, it was too expensive to maintain and closed in 1976. Tennessee Military Institute (1909), originally founded as Sweetwater Military College (1874), ceased functioning as a military school in 1975 and was renamed TMI Academy. The school, however, would close for good in 1988.

Between the closing of Brown Military Academy in 1958 and the closing of San Diego Military Academy in 1979, roughly thirty notable military schools closed across the nation. In the San Diego area, the Elsinore Naval and Military School closed in 1977. By this point, the Army and Navy Academy was now the only military high school and one of three military schools left in the state; all were located in Southern California. The other two survivors were the junior schools St. Catherine's Academy in Anaheim and Southern California Military Academy in Long Beach.

Chapter 15

Strange Sense of Nostalgia

Colonel Cline walked along the old familiar grounds of the Army and Navy Academy. He recalled every crack and pothole. The old chipped paint, the creaking buildings—everything looked the same as Cline had remembered. Although this was a nice trip down memory lane, this was a bit disconcerting to the colonel. Cline retired from his Senior Army Instructor position at the Academy in 1968, but he would nevertheless visit the campus whenever he got a chance. On this day, however, the visit was not for leisure. He was back due to a call he received from Colonel Atkinson.

It was 1982 when his old friend gave him a call. Cline, having since been promoted, automatically jumped at the chance to visit the Army and Navy Academy again, a place he always considered home. Now on campus once more, Cline noticed how ANA had not changed in the fifteen years since he had left.

Cline was shocked to see Atkinson in bad shape. Atkinson had just come back from a trip to Florida, where he was mugged and severely injured. As much as he wanted to lead his old school, he knew he just did not have the health or energy left. Atkinson requested that his old friend take care of the Academy when he was gone, asking Cline if he would consider taking the position of vice-president to support President Brookhart, who was failing in health.

Donald H. Cline

In October 1982, Gilbert N. Brookhart passed away, the second ANA president to die while in office. He had served as Headmaster since 1974. The Board made Fred Rowbotham, business manager and Board member, the next president.

In 1983, Colonel Atkinson passed away, having served fifty-nine years at the Academy since being brought on as bandmaster in 1924. The old guardhouse near the front of the campus was converted into a museum and named in honor of Atkinson. While Davis would be remembered as the founder, Atkinson would be remembered in higher regard as the figure who most shaped and defined the Army and Navy Academy.

Cline mourned the passing of his friend and wanted to honor Atkinson's final wish by serving as the next president of the Academy. In July 1983, Cline retired from active duty in the U.S. Army to dedicate more time to the Academy. In August, the Board finally agreed to make him the Academy's ninth president.

Now in his new position, Cline sought to improve the school rather than keep it a nostalgic relic from the 1960s. After making some observations, Cline realized why the campus had not changed in the last decade. Since

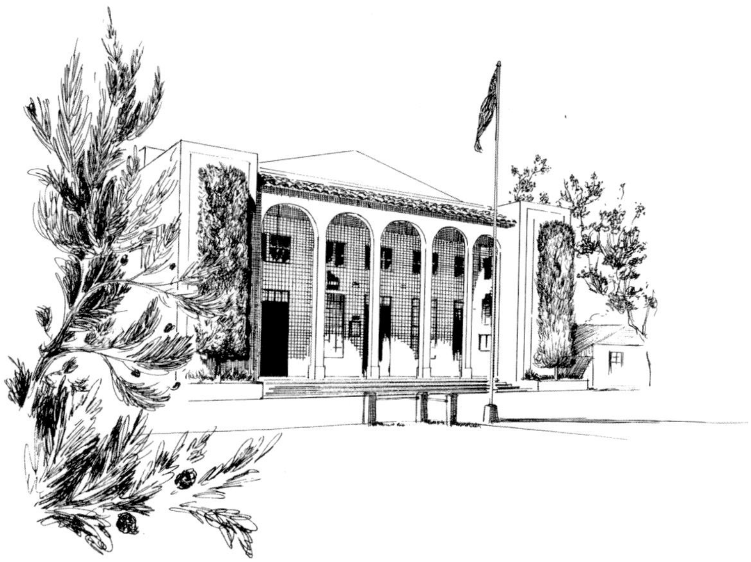

Copy of an ink illustration of Davis Hall during the 1980s. *Author's collection.*

Atkinson's retirement, the Academy had fallen under the control of a small three-man Board of Directors: Thomas Sefton, president of the San Diego Trust and Savings Bank; the aforementioned Fred Rowbotham; and Vroman J. Dorman, the old attorney of the Andersons and the Academy.

Dorman had been with the Academy for many years. He had worked with Davis since the Academy's relocation to Carlsbad. He was a cunning man with financial expertise. While alive, Atkinson had kept Dorman in check. Afterward, Dorman and the two other civilian board members ran ANA with unlimited power. Cline knew that nothing would improve, as these civilians alone knew nothing of military school operations. Cline made himself the fourth member of the Board; however, that made little difference. Voting was always three to one, with the civilians always winning. Whenever Cline requested funds for the improvement and betterment of the Academy, he was always turned away.

The Military Schools of Glendora

William G. Harding (known to cadets as "Old Man Harding") moved to Glendora in 1934 after taking a job at Carlin Military Academy. Carlin was a small, deteriorating school owned by George D. Whitcomb. Harding was never paid for his work at Carlin. When he approached Whitcomb for his paychecks, Harding ended up buying the school instead and reopened it as Harding Military Academy. The school was operated by three generations of the Harding family. The school's highest enrollment topped out around 150 in the late 1950s.

Captain Stoll served as Commandant at Harding Military Academy in Glendora. Harding was a military school for young boys from primary school age to ninth grade. Unlike most military schools, which aligned themselves with the JROTC (which operates only with high schools), Harding was a part of the California Cadet Corps system.

Stoll's office was in the armory of Harding. Cadets who attended recalled that being called into Stoll's office was the most dreaded thing to occur to any cadet, yet many remembered him fondly. After facing a decade of declining enrollment, Harding closed its doors on July 6, 1972. By then, Stoll had relocated to ANA, where he continued to serve as a Commandant to the Lower School cadets. He dedicated his entire life to teaching, coaching and counseling younger cadets. In 1980, the Junior School drill team was named

Stoll's Guard in his honor, and in 1989, for his service to the Academy, he was memorialized with a statue that stands on the Senior Lawn.

Along with Harding, the other military school in Glendora was the relocated Brown Military Academy, which operated on the former campus of Brown School for Girls. As Brown was a high school and Harding was a lower school, it was common to see students who graduated from Harding continue on to Brown. Some Glendora boys spent their entire childhood in military schools, starting in Harding and finishing in Brown.

Brown Military Academy never rose to the same prestige it once held while in Pacific Beach. Most staff and cadets would continue on to Louis Bitterlin's San Diego Military Academy, taking with them most of the DNA that made up Brown. One alumnus who transferred from Pacific Beach to the Glendora campus recalled, "Glendora, which was a much younger campus, but to us, it was a girl's school—like aw man, a girls' school." The new school was small, with enrollment never exceeding more than 150 cadets.

The new campus continued running a JROTC program. There was no need for a junior school, as the nearby Brown-owned Southern California Military Academy occupied that role. Most of Glendora's enrollment came from SCMA. One notable event that occurred happened in 1963, when former president Eisenhower visited campus for the dedication of the newly built barracks, General Eisenhower Hall.

Brown in Glendora closed in 1966; the campus grounds were bought by Azusa Pacific University. A marker still stands on the site of the former Brown campus in Glendora. The Brown alumni from both the Pacific Beach and Glendora campuses formed an alumni association, along with the alumni from Southern California Military Academy. However, the Pacific Beach alumni would often tease the Glendora alumni as having attended the "girls' school." When Brown alumni gather for a reunion at the Glendora campus, most Pacific Beach graduates choose not to attend, considering Glendora a completely separate school.

The Southern California Military Academy

The last John Brown campus to remain in California was the Southern California Military Academy, which was founded by a group of Long Beach businessmen in 1924 as the South Western Military Academy. The following year, the school was purchased by Mr. and Mrs. Wilbur J. Watkins, who

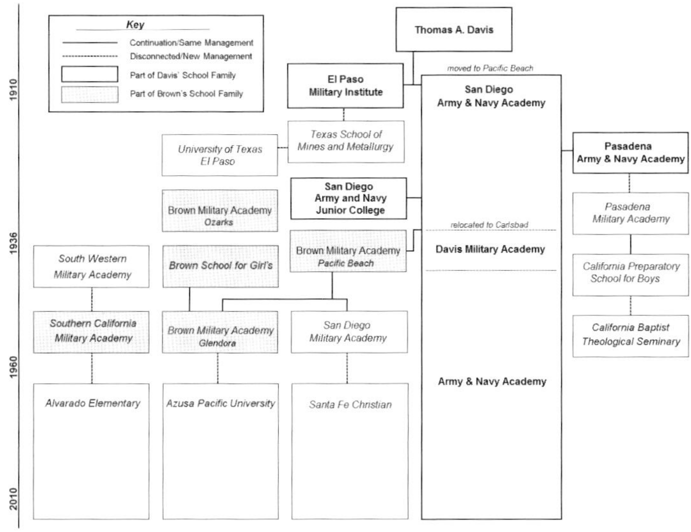

Family tree of the Thomas A. Davis/Brown military schools and their current fates. *Chart by Alexander Mui.*

operated the campus as a lower- and high school–level military academy, until February 1947, when the property was purchased by John Brown. Brown turned SCMA into a lower school and intended his new school to serve as a feeder for his main military school on the Pacific Beach campus, much like what Davis intended with the Pasadena Army and Navy Academy. Cadets who graduated from SCMA after the ninth grade were strongly recommended to attend Brown Military Academy (when the campus was still in Pacific Beach) to conclude their high school education; ironically, most went on to finish at the geographically closer Army and Navy Academy.

After John Brown Schools dissolved its California holdings in 1979, SCMA became a private corporation. The school remained a lower school and retained its strong emphasis on Christian education common with Brown schools. The year before, in 1978, San Diego Military Academy closed its doors. By this point, the only other military school left in the Golden State was St. Catherine's Academy in Anaheim—like SCMA a primarily Christian school, with lower schools not governed by the ROTC.

Tiny SCMA occupied a five-acre campus in Signal Hill, a city within Long Beach. The school's motto was "Character before Career," which was a continuation of the motto of the John Brown Schools. At one point, the school did peak at around four hundred cadets, from kindergarten to ninth grade. However, the school would not be spared from the economic recession of the 1980s. Signal Hill decided to close the school, and in 1987, the campus was completely demolished. When the alumni from BMA of Pacific Beach and Glendora formed the Brown Military Academy Alumni Association, the alumni from SCMA were initially included. However, in recent years, the SCMA alumni have separated and led an effort to piece together their school's history. A bronze plaque was dedicated on the former school grounds, now Alvarado Elementary School.

The military schools of California once numbered in the fifties. Now, the Army and Navy Academy is not just the only military high school left in the state but also the sole survivor west of the Rocky Mountains.

ASSOCIATION OF MILITARY SCHOOLS AND COLLEGES OF THE UNITED STATES

Cline knew that ANA was on its last legs. As military academies were nearly obsolete, she would fall to the same fate as the countless other military schools now long forgotten unless she could evolve. Shortly after his arrival in the fall of 1983, Cline began attending meetings of the National Association of Independent Schools and the Association of Military Schools and Colleges of the United States (AMSCUS) for support in keeping the Academy stable. This organization once had prominent Academy members like Colonel Atkinson among its leadership ranks. The AMSCUS, founded in 1914, was an especially important group in helping preserve the dwindling military school community in the post–Vietnam War, post-Watergate cultural landscape. This nonprofit organization is made up of member schools that are approved by the Department of Defense and maintain academic accreditation. Only military schools that met its high standards were welcomed as members of the organization.

The AMCSUS has somewhat become a self-help focus group, where representatives from the few remaining military schools would share information with one another to help keep their various schools afloat. By attending meetings, Cline learned how other military schools were being run and what improvements could be done with the ANA. After attending these meetings, he would present his findings to the Board along with plans on school improvement. Cline pointed out that almost all peer schools had Boards consisting of eight to sixteen members, which normally would include prominent individuals from the educational institutions and

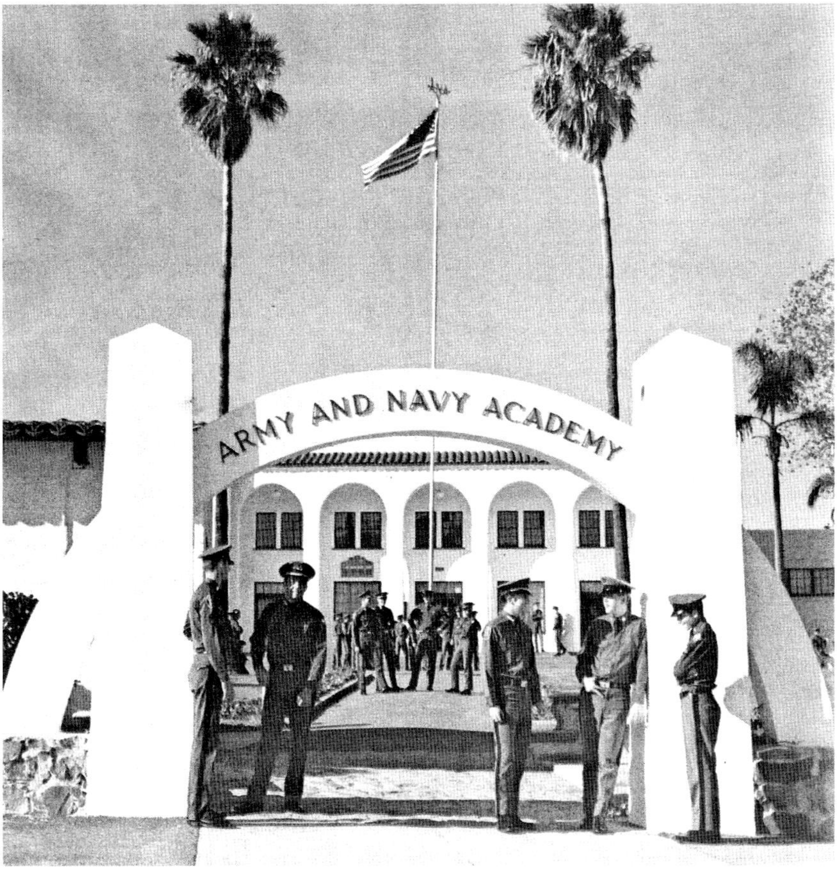

Cadets around the Army and Navy Academy's archway. *Author's collection.*

organizations in the business world. He also pointed out that a general or admiral would bring prestige to the name "Army and Navy Academy." But the civilian members ignored his ideas, and Cline would always leave the meetings stunned by their responses.

BATTLE WITH THE BOARD

When Cline tried to present his ideas to the Board of Trustees, they were frequently rejected. This created frustration among the administration and led to improvements, if any, being done at a snail's pace. Although his ideas were consistently rejected, Cline still wanted to find a way to ensure that the cadets received the highest of college preparatory education. This meant that he had to retain and recruit a good staff and improve basic educational facilities. One problem he had to face was the fact that the salaries for the faculty were less than half of the starting salaries of public school teachers in the local area. Another example that showed how the Academy was out of date was the fact that cadets were still using antique manual typewriters, as opposed to computers, which were becoming the norm at every school during this period.

With the Board being a dead end, Cline turned to the public for help. He presented his ideas to parents and alumni, who in turn provided their support. During his reign, salaries of the faculty increased by 80 percent, and staff increased 50 percent. This attracted many quality instructors and staff members. Yet such actions incurred the wrath of the Board. Along with increasing salaries, Cline also renovated the old campus. A computer lab was constructed in the old storage room, and the latest model computers replaced the old typewriters. The gymnasium floor got a modern makeover, while the stage curtains were replaced. New lights and sound systems were installed along with the soundproofing of the gym. The basement of the gym was also remodeled. Originally, the basement was divided by a wall. On the right was the indoor rifle range, and on the left was another storage room. With permission from the city, the wall was knocked down, with care taken not to collapse the gym. The basement was made into the new weight room, with the indoor rifle range converted into a wrestling room. The old weight room was turned into the field house. Before the field house was obtained by the Academy, it was part of the popular Amigo Café, which provided cadets good off-campus meals. The rifle range was moved outdoors by the tennis court. The beachfront Recreation Hall was also remodeled, along with a faculty lounge and teacher's dining room being added. Across the street, the football field gained a new press box, scoreboard, football and soccer goals, baseball dugouts and a baseball screen. The Board was not happy.

The civilian members of the Board took issue with Cline's decision, which led to awkward Board meetings. The "History on Select Presidents of the Army & Navy Academy" stated:

> *Board Meetings usually took place in Thomas Sefton's office in San Diego with the three civilian Board members seated comfortably on an expensive couch, with Colonel Cline seated directly before them in an uncomfortable wooden chair. At*

one point the Board ordered Colonel Cline not to wear his military uniform to Board meetings anymore because it made them feel too intimidated.

One of the biggest challenges faced by Cline was to get the Board to approve his proposed establishment of fundraising programs. The Board reluctantly agreed but said that it must operate without the Academy's support. No money was to be provided for the program from the Academy's budget, and the Board members would not participate in fundraising efforts. Cline went out and started printing pamphlets and brochures explaining what the Academy was all about. He hoped that this would affect donations from alumni and the surrounding community. Later, he convinced the Board to allow him to hire a development officer. The Board said this would only

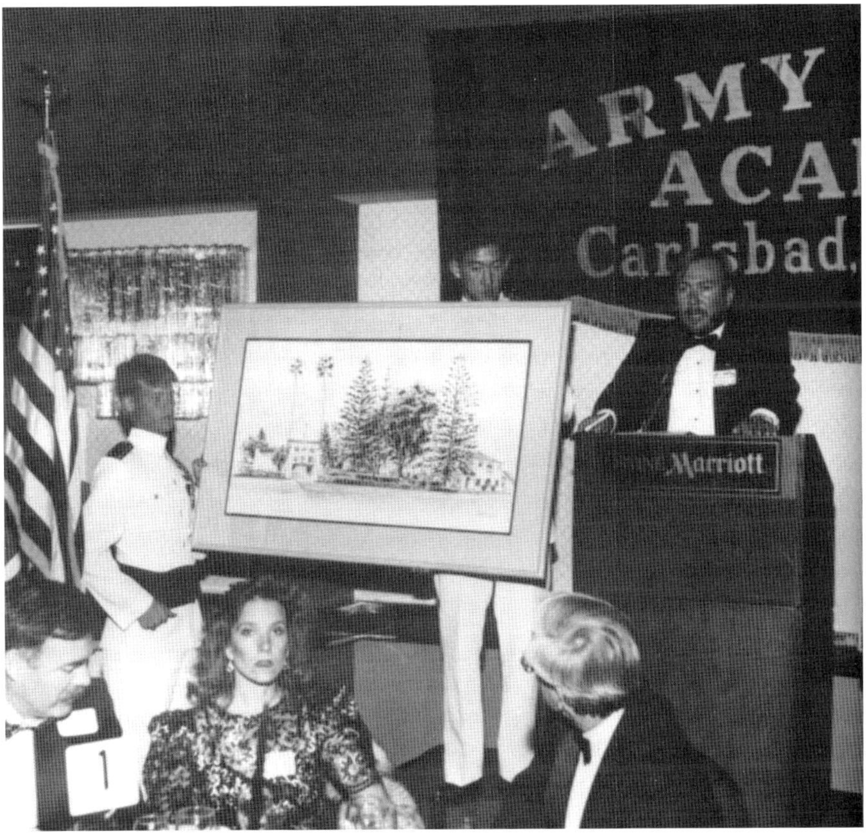

The Salute to Colonel Atty Fundraiser, held by the Alumni Association in 1988 at the Marriott. *Courtesy of ANA Alumni Association.*

be allowed if the Development Program brought in enough money to pay the officer's salary.

In an interview with *The Warrior* in an article titled "Big Dreams for ANA Future," Cline noted, "The development program is to raise money to help cadets to have a better education here at this school, by providing better facilities and tuition assistance, also by providing scholarships for new people to come. We need money to make these dreams come true." Later, the Board eliminated the development officer position. Cline described the news of the Board's decision as a sad day; however, he was grateful that the Patrons Association and Alumni Association stood by him.

Cline continued doing what he could to improve the school on one front while also fighting the negative reputation military schools had during the post-Vietnam era. Despite being a West Point graduate, Colonel Cline worked hard to downplay the military portions if they did not benefit cadets upon graduation. First and foremost, he wanted to make it clear that the Academy is an academic institution that stresses college preparatory studies, not combat skills. In an interview with the *Los Angeles Times* in 1989, Cline remarked, "We're hoping and trying to develop such traits so that the young man can carry them with him into the *civilian* world long after he leaves the academy." Cline placed noticeable emphasis on the word *civilian*. "We believe that it's a good idea to develop certain traits, like self-discipline, at an early age." Cline continued:

> *We do not train mini-Rambos.... We're constantly fighting the misconception that we're grooming killer fighters. Or, people think that just because we have a structured program, that we're a reform school for troubled kids. That's not the case at all. Notice we don't have walls and fences; we have rules and regulations.... We don't make them march because we're trying to wear out shoes. We don't believe in piling punishment upon punishment until a point where a cadet feels there's no way out.*

In 1985, the Army and Navy Academy celebrated its Diamond Jubilee. To honor the occasion, the Class of 1985 donated a large stone sign bearing the name of the Academy, which sits on the Senior Lawn welcoming all to the campus. In 1988, a successful fundraising program was organized by the alumni association. Called "Salute to Colonel Atty," the fundraising was dedicated to Colonel Atkinson and his service to the Academy.

CELEBRITY CADETS

The Army and Navy Academy has seen her fair share of interesting cadets and patrons, from an Arab prince to a trio of Taisahnese royalty and sons of Soviet generals. The diverse pool of individuals who were members of the Warrior Battalion include General William W. Crouch, the former vice chief of staff of the United States Army; Major General Supridee Pravitra of the Royal Thai Army; swimmer Robert Beach; Baja California governor Eugenio Walther; Benihana restauranteur Kevin Aoki; skateboarder Tod Swank; Flogging Molly band member Matt Hensley; actor Marc McClure; and *Rain of Gold* author Victor Villaseñor, among many others. There were even rumors that a young Leonardo DiCaprio had enrolled for a few weeks with the Class of 1992 that began to surface in 2011 after he appeared courtside at a Lakers game wearing an ANA baseball cap. However, in terms of celebrity culture, it was when the Simpsons joined the ANA family that the nation really started talking.

In 1985, it is fair to say that O.J. Simpson was one of the most popular celebrities in American culture. Around this time, he sent his eldest son, Jason, to ANA to be in a distraction-free environment. "I wanted to get him in an environment where school was the focal point. Jason needs that structure," he told the *Los Angeles Times*.

As a young cadet, Jason commented, "I was kind of hesitant because it was a military school….The first two weeks of school were like murder. I was so homesick." Eventually, he settled in and began to enjoy the new environment. "It's a military school in California on the beach. Some of these guys are surfer guys in uniform. I don't even really think of it as a military school."

While the NFL star never pressured his son to join the football team, knowing that he been rejected from the team at his previous school, he was proud when Jason joined the Warriors as a halfback and leading rusher. O.J. came to campus to catch every game, including one in which he witnessed Jason return a kickoff eighty yards for a touchdown. While O.J. never expressed his joy face to face with Jason in fear of him getting "the big head," he enthusiastically told millions of viewers across America during a network broadcast of a *Sunday Night Football* game of his son's touchdown, after which cadets watching yelled, "Jason, did you hear what your dad said on TV?"

"You see a lot of his father in him," Coach Maffucci said. "He plants his right foot just like his father did." Jason Simpson graduated with the Class of 1988, went to college in LA and later started training as a chef.

Unfortunately, around this time, his father would be known for the widely publicized criminal trial in which O.J. was acquitted for the murders of Nicole Brown and Ronald Goldman. As the public started to speculate about the murders, some private investigators began to theorize that Jason was the true murderer, including Detective William Dear, who contacted Jason's ANA classmates on the matter. Dear and others noted that Jason's time at the Academy, which included training in hand-to-hand combat and field knife training, made him the most likely suspect.[22] Back on campus, another legal battle started brewing.

BEGINNING OF THE DARK AGE

On the morning of January 16, 1991, Colonel Cline received an abrupt phone call from Vroman Dorman. Dorman informed the colonel, "A decision had been made to terminate you, Colonel Cline, as President." Dorman continued, stating that "Dr. Conrad Garner and his wife are coming to immediately take over, with a special meeting of the Board of Directors scheduled for January 25, 1991, at 9 a.m. in the offices of Thomas Sefton, at which time formal action to terminate Colonel Cline will be made."

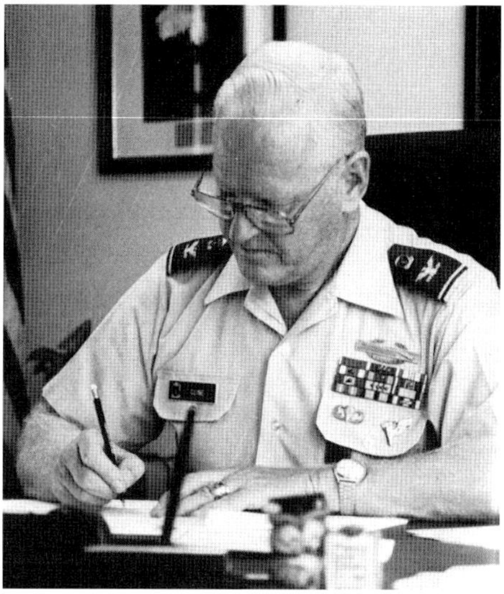

Photograph of Colonel Cline, busy at his desk. *Courtesy of Army and Navy Academy.*

Cline was told that Sefton and Rowbotham agreed with this decision. From the Board's perspective, Cline had undermined its position and drained the school of funds. From Cline's perspective, he did his best to *save* ANA from the fate that had befallen other military schools during this period.

22. "Even some of Jason's former roommates at the Army and Navy Academy emailed me describing Jason…they called him J.J.—While attending the Army and Navy Academy, Jason was trained in hand to hand combat as well as field knife training." From William Dear, *O.J. Is Innocent and I Can Prove It* (N.p.: Skyhorse Publishing, 2012).

With dignity and great love for the cadets and Academy, Colonel Cline graciously stepped down from the role he had served for seven years. Since being hired as the first Senior Army Instructor in November 1966, twenty-five years of service to the Army and Navy Academy concluded when his duties as president ended on January 25, 1991. Cline later said, "I always did, and always will consider it an honor to have served as President at the 'West Point of the West' and will always view the Academy as home." With that, the Academy's Bronze Age gave way to its Dark Age.

Donald Cline returned to his beloved "paradise" only once since his retirement in 1991, nearly two decades later in 2010, for the Army and Navy Academy Centennial Celebration. By then, most of the individuals who had opposed Cline were dead or long gone. He was one of two former presidents present at the event. Cline passed away on June 18, 2014, and was interred at West Point Cemetery on the grounds of his alma mater. Before his passing, he wrote his own memorial on the West Point record:

> *I have decided to accept the responsibility of writing my own Memorial Article, thus not burdening anyone else with the chore. After all, who knows me and my life better than I do? I've been privileged to have made many friends, served with outstanding companions, and enjoyed a good life. I guess all good things must come to an end.*

CHAPTER 16

THE YEAR OF FIVE PRESIDENTS

HAD DONALD CLINE not been forced out by Dorman's Board, Army and Navy Academy history might now refer to a Cline Era of progress and modernization. However, history is never written as intended, and what appeared to be a ray of hope for the Army and Navy Academy during turbulent times was snuffed out as Vroman Dorman increased his power over all Academy affairs. The period that followed was one of uncertainty, which could very well be retroactively called the Army and Navy Academy's Dark Age. There are two reasons for this: First, this marked an era when revenue, accreditation and enrollments were at an all-time low. Second, while recent, it is a period of history that is barely touched on, often skirted over in official publications dealing with the school's history. However, as with most histories, there are numerous sides and stories; in the age of public access and mass communication, this means that rarely do such stories stay hidden. However, the history of the Army and Navy Academy is not about any lone individuals or moments, but of a broader institution that has been shaped by countless individuals yet, in some sense, remains the same at the core. And as the Academy tumbles out of a period of military school closures and the removal of a progressive president, she now must combat a period of turmoil and continue to march into a new millennium with her head held high.

The Presidents of 1991

The poorly planned termination of Cline meant that a new president had to be found quickly. Dr. Conrad Garner and his wife came from Wilseyville, California, to the campus when the Board appointed him as the new president on January 25. He was asked to quickly create a plan designed to slash costs dramatically. This would include letting go of one-third of the faculty and staff.

Mrs. Garner lived on campus, while Dr. Garner would trek back and forth from their old house until they could completely move in all their belongings. On April 10, 1991, the two showed up with their moving trucks at the President's Residence. But within a few months of moving in, in the middle of the night, they mysteriously repacked their bags and drove away. Maffucci likes to jokingly remark that it was the cadets who scared them away. They never returned to the Academy. His term lasted less than four months.

In a rush to find another president, new Board member Johnnie Crean suggested they appoint the man he served under as equipment manager, the popular athletic director, Coach John Maffucci. Maffucci just finished with an 8-0 football season in 1990, a record for the team, which resulted in the Academy playing in the Division I Championship game. With his popularity after the impressive 1990 football season and his years of service to the Academy, many saw Maffucci was an appropriate choice for the role of president.

Maffucci was made acting president of the Academy in April 1991, remaining in that role until the end of summer. During that time, he still continued to coach his sports teams and run Camp Pacific. For his dedication to the Academy, in October 28, 1994, the football field was named Coach Maffucci Field, and Carlsbad declared that day as John Maffucci Day. He is also honored in the San Diego Hall of Fame.

For the 1991–92 school year, the Board brought on Dr. Michael Wallgren to serve as president. However, this did not save the Academy from her rapid decline post-Cline. That year, enrollment fell to 143 cadets, the lowest in history since the relocation to Carlsbad.

The Crean Foundation

John and Donna Crean, a successful couple who founded Fleetwood Enterprises, a Fortune 500 company, had four children—all their sons attended the Army and Navy Academy. In their home of Orange County, they are known for their philanthropy and generous donations. Their Crean Foundation donated millions to notable causes, including a $10 million gift to Chapman University and millions a year to charities. The Creans were honored with the tenth star on the Anaheim Walk of Stars, right alongside Walt Disney, for their dedication to the community. As the sons from the Crean family attended the Academy as cadets in the 1960s, just as their sons would later, the Crean Foundation became the school's largest benefactor during this period.

In 1991, one of their sons, Johnnie Crean, the same cadet who served as equipment manager for the football team, would serve on the Board of Directors, brought on as a suggestion by Dorman, as he believed having an alumnus from a family of wealthy philanthropists would benefit the Academy. During the meeting at which Johnnie was brought on, he noted that the Board members would not speak to one another; they all agreed to vote him onboard and then abruptly departed. Commenting on this meeting, Johnnie noted, "They hated each other, all of them; they would never get to a place where they would have all three votes. Until they elected me. When they elected me, I knew it was not running well. I ended up running the operations committee. I ended up with all the responsibility; it was pretty daunting."

In 1992, the Crean Foundation made a generous donation to the Academy, a gift of $1.5 million, which was used to fund the construction of two new dorms. Bain Hall was torn down to make room for the new John C. and Donna Crean Hall and the William C. Atkinson Hall, which were completed in 1994. Upon completion, these dorms became the largest structures on campus.

THE END OF DORMAN

As Vroman Dorman's health declined, he delegated more power to Johnnie Crean, who brought on his classmates to serve as new board members. On December 21, 1993, Dorman passed away. Today, he is remembered by the Academy in much the same way America remembers Benedict Arnold. Like Arnold's victory at Saratoga that allowed America to win its independence, Dorman was responsible for keeping the Academy afloat during critical times. His experience with handling money and financial reasoning grounded the idealistic natures of Davis and Atkinson. However, Dorman was also blamed for failing to direct finances where needed, which resulted in the deterioration of the Academy back in the 1980s.

At the time of his passing, Dorman was already memorialized on campus with an old dormitory. In 1957, an uneven dorm building with an elevated kink in the middle was named after him. Dorman Hall was built on a sloping hill, which resulted in its slightly odd, crooked appearance. Half the structure was elevated higher, and then the building abruptly dropped a foot and sunk into the ground. After Dorman's passing, the High School drill team, previously called Milton's Guard, was renamed Dorman's Guard in his honor.

In modern times, Dorman is remembered by ANA in a more negative light. He is often remembered for selling the beachfront athletic and parade fields, known as the Point. Old rumors have it that the money from the sale of this land have vanished without a trace.[23]

Those rare individuals who speak highly of Dorman believe he does not get the credit he is due. His supporters note it was Dorman who was responsible for convincing the Andersons to gift their land to the Academy and claim the reason he is hated was because he was the individual responsible for Thomas A. Davis leaving Carlsbad half a century ago. These claims, while they might hold some truth, are difficult to completely verify either way.

THE RELUCTANT PRESIDENT

When Wallgren left in 1992, Crean asked Dr. Fred Heinle to serve as the next president. Heinle served as principal of Orange Glen High School in Escondido after the school received two years of probation after its failure in the accreditation process. He worked quickly, and within two years, Orange Glen received the highest accreditation possible, named a Distinguished High School.

Heinle was starting his own private school with his wife. He was asked to help save ANA's accreditation with the Western Association of School and Colleges, which put the school on probation for two years. Heinle was initially not interested, as he was not too fond of working for a military school.

23. I cannot verify this claim. However, among alumni and old staff I have spoken with, this is a story commonly accepted as fact.

Heinle had been a private in a post–World War II German army, an army that did not think highly of its profession considering the scars left by its predecessors. Heinle would grow up strongly disliking the military; however, Heinle agreed to examine ANA before making any decision. Upon observing the school, he concluded that military and disciplinary programs overshadowed the academic portion. Heinle saw that students were doing marching and drill instead of learning, and disciplinary punishments were taking up too much time. Cadets waiting in line for food had to do pushups, causing them to eat with dirty hands, which he concluded led to increased sickness. Heinle told the Board that he would take the job only if he could take charge of the whole school, including the military programs. He wanted the cadets to stay in the classrooms and eliminate physical punishment. The Board agreed with his demands, and he became the president on July 13, 1992.

Heinle improved ANA, just as he did with Orange Glen High School. In 1995, when WASC visited ANA, it reported outstanding change in every area. WASC give the school the highest reward, a full six-year accreditation. Only 5 percent of all schools receive this high rating. When WASC visited in 2001 and 2006, ANA continued to maintain its high rating.

THE GREAT FIRE OF 1999

In the early hours of March 2, 1999, a cadet living in Dorman Hall went to bed, forgetting to blow out a candle he had burning. The candle was placed on top of a wooden footlocker, which burst into flames when the candle fell over. Cadet Joon Kang, who was near the pay phones, saw flames erupt from the room and quickly tried to put it out.[24] TAC Officers McCastle and Richardson quickly began evacuating the building, while senior cadets woke up the entire Corps. Within a few minutes, they had organized an accountability formation in front of Crean and Atkinson Halls.

Fire spread rapidly, fueled by the old rotting wood structure of Dorman Hall. Finally, the fire truck arrived, and the flames were put out, but Dorman

24. The 1999 yearbook noted that Dorman Hall erupted into flames when a middle school cadet's candle fell over. However, I have been told various stories, including that a portable heater (which I originally had considered the official version of the story) or an electric hotplate or hot pot (which was the original story I was told) sparked the fire. As the yearbook contains the earliest printed source, this is the version of events I tell.

The Dorman Hall Disaster

Twenty-foot flames engulfed Dorman Hall, sparing nothing in their path.

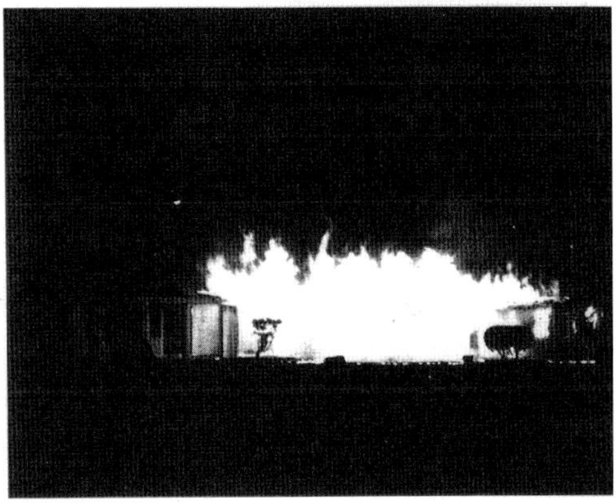

On one dark morning, March 2, 1999, disaster struck at ANA. It was 1:40 in the morning when a young, middle school cadet's candle fell over during his sleep into his foot locker and sparked a fire. The foot locker was by his bed, and in seconds the bed caught fire. Joon Kang saw the fire brewing in the cadet's room from the phones and tried to put it out, but the fire extinguisher and his efforts could not tame the flames. Instinctively, the senior and Capt. McCastle began the evacuation of the hall and surrounding areas, all within a short period of time.

While all this was happening the cadet corps watched in awe and terror, thinking to themselves "how could this happen to us?"

The Senior leadership stepped in after a few minutes and started the ball rolling to evacuate the rest of the halls and had an accountability formation in front of Crean and Atkinson Hall.

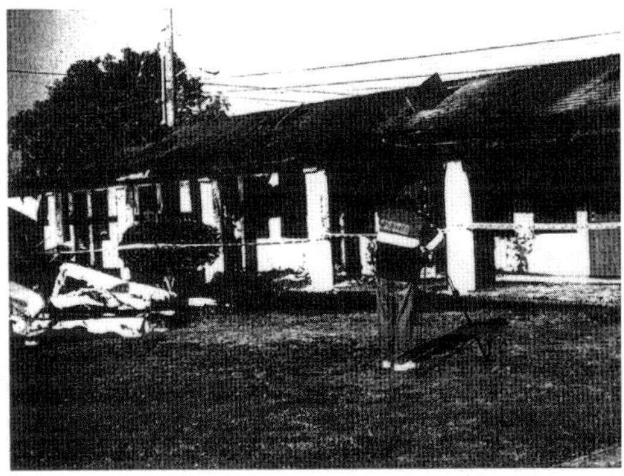

Fortunately no one was hurt, but those cadets who lived in West Dorman Hall were left without a room to sleep in. Cadets from all over campus helped to convert the gym into a shelter for those cadets who lost their rooms during the fire. Capt. Donovan, assisting Col. Miller, brought out pillows and blankets from the uniform store and mattresses from storage.

The morning after the fire, a special chapel was held to discuss the tragedy. Cadets were given an opportunity to discuss the tragedy and were honored for their heroism. After the chapel, cadets attended home room to discuss the incident. Mentor teachers acquired telephone numbers so parents could be contacted about the incident. Immediately after home room, cadets returned to their rooms for an opportunity to catch up on sleep which was lost.

As a result of the fire, the building plans have changed. The new eighty-bed dorm will be built where Dorman Hall originally stood. The Academy has had an opportunity to review the emergency procedures and lessons have been learned from the fire in Dorman Hall.

Dorman Hall after the fire while news crews videotape the aftermath.

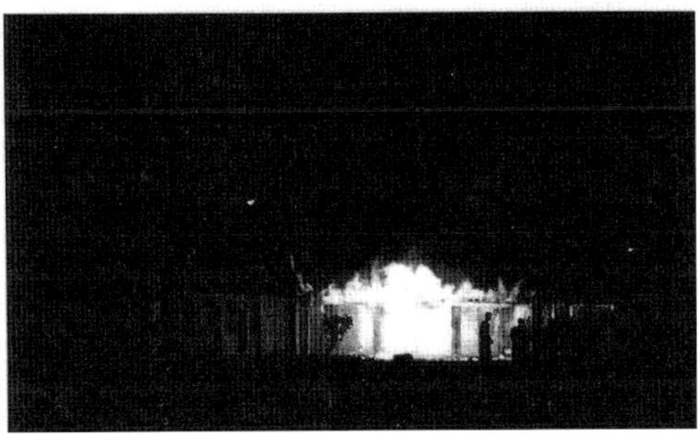

Cadets standing in front of Dorman Hall watching it burn to the ground.

A page from the 1999 yearbook covering the Dorman Hall fire on March 2, 1999. *Courtesy of Atkinson Museum.*

Hall was beyond saving. Luckily, no one was injured. Cadets who lost their rooms had nowhere to sleep. The gym was converted into a shelter for the night, while Colonel Miller and others grabbed pillows and blankets from the uniform store and spare mattresses from storage for cadets to sleep on.

The following morning, a special chapel was held for cadets to discuss the tragic events of the previous night. Afterward, cadets were allowed to return to their rooms to catch up on lost sleep. Captain McCastle, Captain Richardson and nurse Linda Stewart were honored at the next parade, along with the fire crew, to whom Colonel Miller gave awards.

Dorman Hall was gone. A new building was constructed in its place, a large modern dormitory named Anderson Hall to honor the benefactors for whom Dorman once worked. Shortly afterward, Dorman's Guard was renamed Davis' Guard in honor of the Academy's founder. The last remains of Vromon Dorman were now gone. His memory was quickly forgotten. Today, his name is only heard in whispers by the old guard.

THE END OF THE PRESIDENCY

On June 21, 1995, the office of president was split into two administrative positions, Headmaster and Superintendent. Heinle's title was changed to Headmaster, and Lieutenant Colonel Stephen A. Miller was named Superintendent. The Headmaster and Superintendent would work together, filling the role once occupied by the president. For three years and four months, there was officially no president in Academy records.

As a common repetition in the long history of the Army and Navy Academy, there are the occasions when the Board acts as an inhibitor to the growth of the Academy. The 1935 Board was responsible for bringing an end to the San Diego Army and Navy Academy of Pacific Beach. Dorman's Board halted any progress Colonel Cline tried to enact. This new Board, controlled by the money-oriented CEO and his classmates, appeared to threaten the school's nonprofit status. However, this time, the faculty, staff, parents and alumni of the Army and Navy Academy united to reform this Board of Directors and forced it to reinstate the position of president. Through its actions, Superintendent Colonel Miller was named the fourteenth president of the Army and Navy Academy in 1998.

Eventually, the Board of Directors was removed, and a new Board of Trustees made up of alumni, parents and prominent community leaders

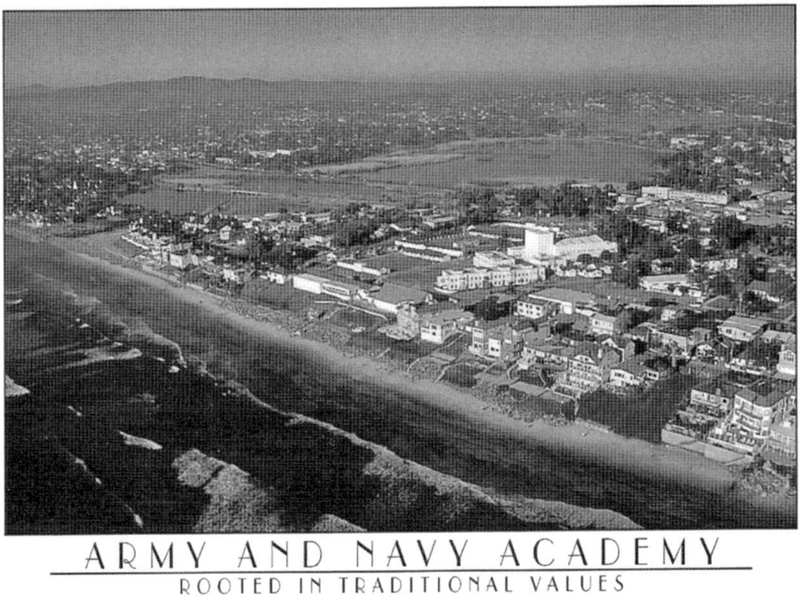

Academy postcard from the 1990s showing an aerial view of the campus from the west. *Author's collection.*

was formed. New rules were put in place to prevent another individual like Dorman from taking over, including making the Academy's president a nonvoting member along with a faculty leader representing the academic staff. Guidelines were also put into place to prevent alumni from any single class serving on the Board at the same time. With that, the Army and Navy Academy left one era and entered into a new millennium.

PART VI

THE ACADEMY'S RENAISSANCE

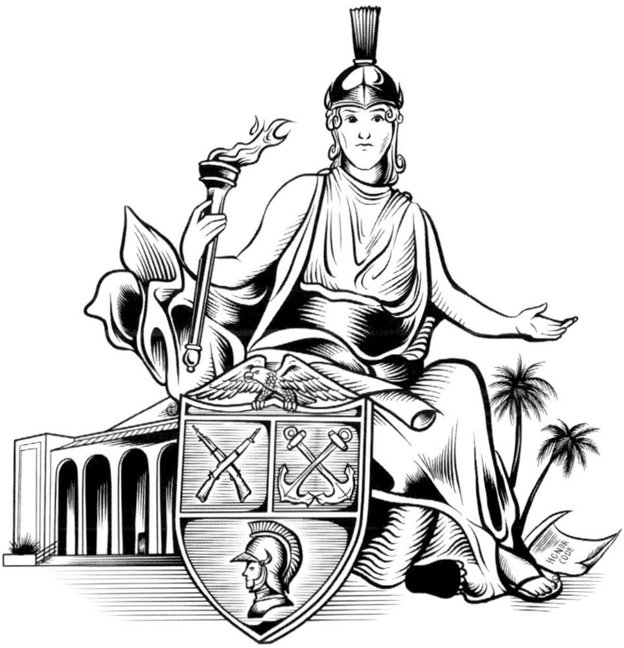

The Academy's Renaissance. Artwork by Alexander Mui.

Chapter 17
Into a New Millennium

It was the morning of September 11, 2001. In New York City, an explosion was heard from above. Debris rained from the tower over people walking on the streets below. They were frozen by the shock. They looked up and horror filled their eyes. It was 0546 Pacific Time, and the Army and Navy Academy Corps was fast asleep. Little did the cadets know what had occurred in New York City at that moment. By 0603, another hijacked plane had hit the second tower of the World Trade Center. By 0637, a third hijacked plane had been flown into the Pentagon. Colonel Miller was out and ordered all cadets to prepare for safety procedures.

ANA was completely in the dark. All that anyone on campus knew was that terrorists were flying hijacked planes into major American landmarks and military locations. There was a fear that California would be a primary target for these hijackers, especially Southern California. The area ANA called home was surrounded by major airports like LAX, John Wayne Airport and San Diego Airport. What was even more disconcerting was the Academy's proximate location to Camp Pendleton Marine Corps Base.

Military bases were believed to be targets of the terrorists. With the Academy being so close to Camp Pendleton, there was a fear that with the cadets in uniform, from the air the campus might be confused for the Marine Corps base. All cadets were immediately ordered to get out of uniform and into civilian clothing, the first time in the Academy's history this was allowed.

New fear erupted when the pilot of Korean Air Flight 85, which was headed to the West Coast, entered the letters "HJK" into an airline text message at 0808 Pacific Time. This was the code for hijacked. By now, nearly every flight within or headed to the United States was redirected or grounded, all but Flight 85. At 1024, the pilots of Flight 85 then changed their transponder signal to the four-digit universal code for hijacked.

The plane was determined not to be hijacked, and the pilots reported that they were ordered to change their transponder signal in an error of translation. Even with all planes headed to the West Coast being grounded, ANA remained enveloped in a cloud of fear due to the near communication blackout.

THE RISE OF JROTC DOMINANCE

A century ago, great wars gave birth to a large array of military schools. This changed after the Vietnam War. The War on Terror proved no different. However, interest in military-style education was rekindled. No new military schools in the traditional sense were created during this period; however, already existing military schools saw a spike in enrollment, and JROTC units began popping up across the nation. And why should there be a need for new military schools? In a day when airfare was readily accessible, students from all over the world can easily fly to any of the already established military schools of their choosing. The days when there was a need to create a military school for a particular area have long gone away.

The only "military schools" to be formed were public disciplinary day schools and charter schools that made JROTC or a similar cadet program mandatory for all students. Examples include the Oakland Military Academy (2001) and Delaware Military Academy (2003). However, these pseudo-military schools were hardly like the military academies of the previous eras; they are no different than any other regular public school that operated a JROTC program.[25] Eagle Military Academy (2002)—founded by Colonel Nathaniel Green in Summerville, South Carolina, as a Christian disciplinary school with

25. When I mention traditional military schools, I mean high schools styled after the federal military academies that house boarding cadets and operate a college preparatory program with its own distinct culture, tradition and structure. These do include institutions founded as religious military schools if they have shifted from being fundamentally religious to focusing on character development and leadership training. Thus, I don't include reform schools or public day schools supplanted with a mandatory cadet or scout program in this definition. The AMCSUS, the governing body of military schools, does not include these schools as members.

only eight cadets—is the closest establishment of the new millennium founded in the tradition of the military schools of old, but with only three full-time employees and a dismal corps of thirty, it closed within a few years.[26]

During the course of my research, I have come to the conclusion that the JROTC program was a major factor in making military schools nearly obsolete. The federally sponsored Junior Reserved Officers' Training Corps was expanded in 1964 with the Vitalization Act, which made establishing a unit much easier for every school across the country. During this time, the creation of new military schools conversely stopped. In fact, since the establishment of the Florida Air Academy (1961) and Marine Military Academy (1965), no currently surviving military school has been founded.

As of 2006, there were a total of 3,230 JROTC Units across the country; the allotted cap is 3,500, an amount that vastly outnumbers the American military schools that were ever in existence. If parents want a military-structured leadership education for their children, they need go no further than their nearest public school JROTC Unit or Boy Scout Troop. For the same feel of a traditional military academy, all a parent would have to do was send their child to the nearest private boarding school with an operating JROTC program.

It could be argued that the parents who send their children to traditional military schools do so out of a sense of nostalgic tradition, as it is much more costly to do so as opposed to the free JROTC program. The struggle to just stay open is quite the tall order for a military school, thus tuition rates are normally much higher than at regular boarding schools. On the other hand, the JROTC program for a regular school is much easier to operate and benefits from government funding, which covers the cost of military instructor salaries, equipment, uniforms and textbooks.

At this time, nearly all surviving military schools have now adopted the ROTC program. All these military schools (with the exception of one or two unaccredited outcasts) are members of the Association of Military Colleges and Schools of the United States (AMCSUS). Around this time, the AMCSUS listed roughly forty pre-college military schools in its ranks, all but one (a middle school not eligible) members within the JROTC program. In a manner of speaking, the distinctiveness of these traditional schools has been dampened as a result. The educational structure of these military schools follows the same specific mandated ROTC guidelines. A JROTC Unit from

---

26. Eagle Military Academy was the newest military school listed on the AMCSUS database; however, virtually no information on this school exists, and I discovered that it later was closed.

a military academy is essentially viewed no differently than a JROTC Unit from a public school in the eyes of Cadet Command. This was a worry of Colonel Davis's when he was originally approached to join the ROTC program. He feared that another education system would override the Davis System that he had established.

One military school that tried to keep itself outside the ROTC structure is Forest Hill Military Academy (2008–13), an attempted continuation of the former Millersburg Military Institute (1893–2006), which closed before even finishing its final school year when it was unable to pay employees. In 2008, the campus was purchased by the U.S. Army Cadet Corps, a rival organization similar to the JROTC founded in 1909, when the group learned of the old school being auctioned from the website eBay.[27] Due to low enrollment of about seventy cadets, failing fire code inspections and a major lawsuit, the school was forced to shut down, and the lawsuit forced the U.S. Army Cadet Corps to reorganize as the American Cadet Alliance. Since then, every military high school in the ASMCSUS circle now operates within the ROTC Cadet Command.[28]

The Academy is now listed as the Warrior Battalion of the Fourteenth Brigade, Western Regiment of the ROTC Cadet Command—just one out of the countless JROTC Units in California. However, as the Army and Navy Academy now operates as the only military academy west of the Rocky Mountains, even as part of the JROTC program she stands out during JROTC gatherings and competitions. In local and regional drill competitions, a handicap is normally enforced on the Academy to give public school drill teams a chance to compete.

When I attended the JROTC-sponsored George C. Marshall Leadership Symposium in Virginia along with three other cadets to represent ANA in 2007, I noticed that we were the only cadets not to wear routine army green. Our West Point–style uniform stood out, and we were automatically singled out to lead the other cadets during the symposium. The JROTC cadets we talked to wanted to know more about our time at ANA. The concept of being in uniform and in a military-structured environment

27. While Forest Hill Military Academy did not operate under the JROTC program, it was the flagship school of the U.S. Army Cadet Corps, a national cadet organization like the JROTC, though not sponsored by the government.

28. The only two schools not part of the JROTC program listed as members of ASMCSUS are St. Catherine's Academy and Robert Land Academy. Being a middle school, however, St. Catherine's is not eligible for the JROTC program and in 2010 dropped "Military" from its name. Robert Land Academy operates the Canadian equivalent to the JROTC, the Royal Canadian Army Cadets.

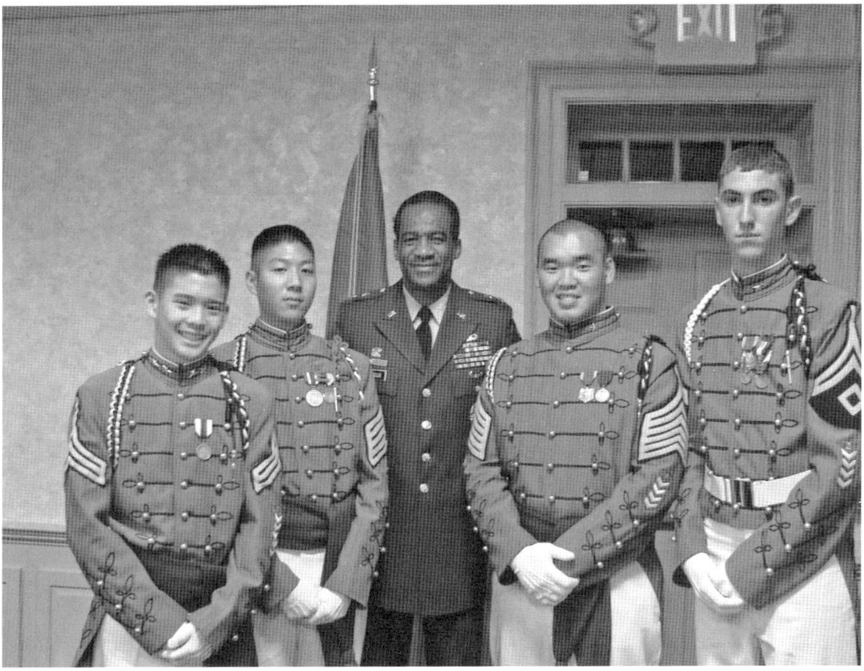

Army and Navy Academy cadets at the 2007 Leadership Symposium with Cadet Command's major general. *From left to right*: Ruangtragool, Mui, Tran and Zoschak, who were the only group of JROTC students in which all four were awarded both the Army Sergeant Major's and Major General's challenge coin. *Author's photo.*

24/7 was alien to them. The symposium was an eye-opening event that perfectly summed up the state of military schools in America. In a sea of hundreds of cadets donning the army green, there were only four who stood out wearing West Point gray.

STEPHEN A. MILLER

For the period between 1995 and 1998, when there was no president at the Academy, that position was split between the Headmaster and Superintendent. The Headmaster was the previous president, Dr. Fred W. Heinle, and the Superintendent was the future president, Lieutenant Colonel Stephen A. Miller.

Lieutenant Colonel Stephen A. Miller joined the Academy during the chaotic year of 1991 as the new Senior Army Instructor. He believed strongly in the system of traditional military schools that fostered a sense of community and strength in the Corps. The SAI deals with the cadets the closest. Knowing this, Miller wanted to use his influence over the Corps to help bring ANA back to the same height of her Golden Age.

For guidance, Miller wanted to draw from ANA's past, hoping to revive old traditions that would enrich the sense of community within the Corps. One of the first traditions Miller wanted to reintroduce is the singing of the alma mater. When he noticed that only a few seniors knew the words to the song, he made the singing of the alma mater a prerequisite for passing leadership training courses. The motto "Warrior Pride" was emphasized and woven into all aspects of Academy life. It was Miller's use of "Warrior Pride" that lives on today as ANA's war cry.

Miller was promoted to Superintendent in 1995 after the dissolution of the president position. His first letter as Superintendent to patrons noted his mission of "providing the best college preparatory program in the finest military tradition." In his capacity as Superintendent, Miller tried to fill the role Colonel Atkinson once held. During the Wednesday Chapels, Miller would adapt many of Atkinson's speeches and related them to current cadet issues. The display of class flags in the chapel was also reinstated. The Middle School drill team, Stoll's Guard, and Mrs. Atkinson's old drama club, Masque and Wig, were also revived. Miller likewise reactivated the National Honor Society and established chapters of the Beta Club and California Scholarship Federation.

Along with reviving old traditions, Miller created new traditions that would be carried on by future generations. Among these was the Eighth Grade Bridging Ceremony. When the Middle School cadets graduated from eighth grade, they would pass under a saber arch made up of the Davis Guard Drill Team. Miller also drafted the Standard Operating Procedures (SOPs) to cover nearly every aspect of cadet drill and campus life by piecing together old guidebooks and listening to alumni descriptions of how things were done in the Atkinson days. This included consistent standards still used today for pass in review parades, new cadet orientation and leadership training. The seeds of the current institution were planted during Miller's term as the Superintendent. All semantics aside, Miller was the president during this era without a president.

THE PRESIDENT RETURNS

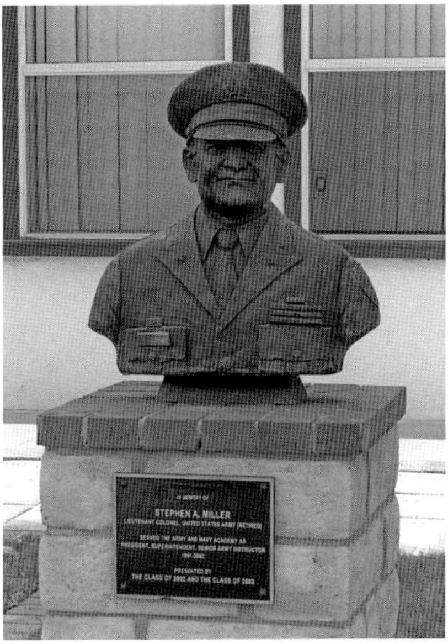

Bronze bust of Miller in front of the Academic Office, donated by the Class of 2002 and Class of 2003. *Author's photo.*

On June 26, 1997, Headmaster Dr. Fred Heinle left the Academy. Heinle served a unique position, coming on as the fourteenth president of the Academy in July 13, 1992, to help fix the Academy's accreditation standing and then serving as the Academic Headmaster for two additional years starting on June 21, 1995. ANA teacher Denis Anderson, retired Marine officer, was promoted as the temporary Headmaster. He filled this role for one school year before Thomas Bloomquist was brought on. Anderson once again would take the role of Headmaster in 2009.

After Heinle left, for one complete school year the Superintendent filled the role of absent president alone in this capacity. On October 24, 1998, the Academy finally reinstated the position of president, and Miller became ANA's fifteenth president. The appointment of Miller was seen as a ray of hope for ANA in a period of turmoil. Normally, the military and academic faculty would struggle for dominance. Miller became a friend to both sides and kept them in harmony, ensuring balance during his term.

During his presidency, Miller guided ANA back to regaining a leadership position with the prestigious AMCSUS, and enrollment swelled from 167 cadets in 1991 to more than 360 in 2001. Colonel Miller also launched a new development campaign that led to many major campus improvements, including the building of the new Anderson Hall, completed in 2002.

Months after the horrors of the 9/11 attacks, the Academy was met with another shock in early 2002 when Miller unexpectedly passed away. At the time of his passing, Miller had served the Academy for more than a decade, as Senior Army Instructor, Superintendent and president. Karen Chemlir Miller, the widow of the late president, noted:

> *He made a point to know the names and family history of each cadet on campus, and would often chat with them in the dining hall. Who can forget trick or treating at the Colonel's House or the senior class meetings and steak dinners? His greatest pleasure was when former cadets and alumni would drop by his office for a visit.*

Old Anderson Hall, which houses the Leadership Education Department, was renamed Miller Hall in honor of the colonel's time leading that department as the Senior Army Instructor. The Classes of 2002 and 2003 dedicated a bust to Colonel Miller that stands outside the Academic Office. As his wife noted, "Colonel Miller would have been humbled to learn the military building bears his name, and a bronze likeness stands among the other notable leaders of ANA."

Chapter 18
Foundations of the Revival

On February 8, 2002, Brigadier General Stephen M. Bliss was named the next president of the ninety-one-year-old Army and Navy Academy. Recently retired from the army, he was the first president in the Academy's history to hold the rank of general. Bliss was born in Philadelphia. He attended the Hill School, which was a small college preparatory school in Pottstown, Pennsylvania. In an interview with *Army Navy* magazine, he noted, "Hill School was invaluable to me when I came to Army Navy and it gave me an excellent framework for understanding how such a school should work."

Along with Colonel Cline and Captain Vestal, General Bliss is one of three graduates from West Point to serve as president of the West Point of the West. General Bliss graduated from the United States Military Academy in 1965.

General Bliss first learned of the Army and Navy Academy from a friend who saw an ad ANA posted on the website of the Military Officers Association of America. The friend informed the general, knowing that he would be interested in this new opportunity.

Stephen M. Bliss

General Bliss arrived at the Academy from Dallas shortly after the passing of Colonel Miller. He noted that he came to ANA not knowing what to expect, as he had no experience in the professional education sector. Despite all the changes Miller put into motion, ANA was still in need of improvements.

When he took on the role of president, Bliss told the cadets and faculty that "anything worth doing is worth doing well." This was the philosophy that would define his presidency. Before he implemented any changes, he took a look at the strong preexisting traits ANA already held. Dedicated employees, quality cadets enrolled and strong alumni and patrons all worked in favor towards improvement. By using these strengths as a solid foundation, Bliss would start to build an even stronger Academy. With the Parents Association and a group of cadet volunteers, Bliss oversaw massive renovations across the campus in 2004, with old dorms being gutted and renovated for cadets. The parents completely refurnished the old Recreation Hall, and the Canteen Snack Bar reopened for the first time in ages.

Under Bliss, the Corps boomed to astronomical levels, surpassing 370 cadets, the largest since moving to Carlsbad. During his term, ANA entered the information age. In 2004, the first ANA website was launched, and the following year, the first modern magazine, *Army Navy*, was published.

As a cadet at West Point, Bliss had seen his share of hazing and bullying. One of his major goals was to eliminate such incidents at ANA. In his first

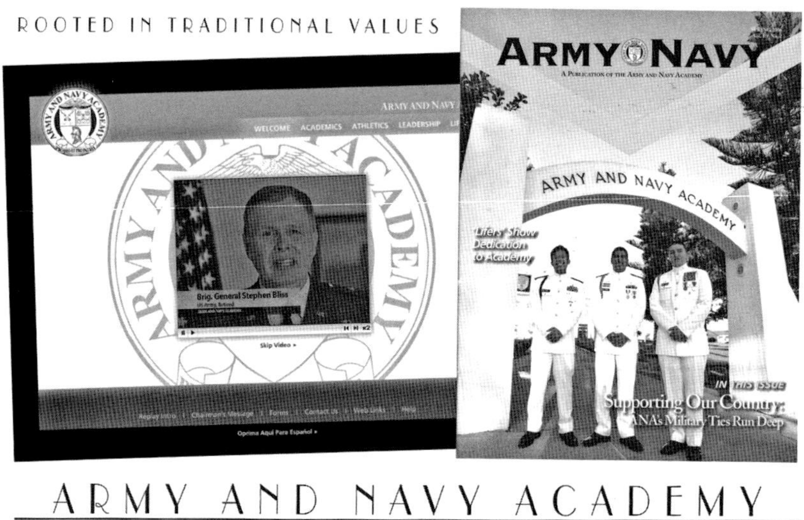

Left: The welcome page of the Academy's website, circa mid-2000s. On the right, the spring 2006 issue of *Army Navy* magazine. *Author's collection.*

year, Bliss enforced a strict policy against such actions by enforcing an honor code that had zero tolerance for such behavior. As a result, the hazing culture on campus became near extinct.

The Honor Code was something highly stressed during this period—a cadet will not lie, cheat or steal or tolerate those who do. Bliss made sure that this code applied to both cadets and employees. More leadership roles and responsibilities were given to the cadets. Cadet leaders began running most day-to-day operations of the Corps, while adult faculty provided direction from a distance. The empowerment of cadet leaders with such responsibilities fostered expedient development of management and leadership skills in ANA cadets, something students in other high schools and JROTC Units could never experience.

In 2005, ANA was made home to the only Vietnam memorial in North County. The North County Vietnam Veteran's Memorial was a project of Jack Frazier and his tenth-grade history class from Carlsbad High School. ANA donated the southwest corner of Maffucci Field to house this new memorial. The memorial was dedicated on Memorial Day in 2008. As the cadet officer on duty, I represented ANA by laying the wreath on the newly christened memorial. The cost of construction was more than $90,000, which was paid for by grants from Carlsbad and San Diego County, along with donations from supportive citizens and the Vietnam Veterans of North County. The memorial is now a notable landmark for people driving along Carlsbad Boulevard.

EDUCATORS OF THE RENAISSANCE

Along with the president, the Headmaster[29] is arguably the most important figure on campus. Peterson, Bain, Ede, Brookhart, Heinle—all figures whose names could mark major eras in ANA's timeline. However, there is one more to include in that list.

During the end of Miller's term, the Academy welcomed Derrick Leasure as the Dean of Academics. Leasure was born in Newark, Ohio, and was educated at Utica High School before attending Ohio State University. He later attended the University of Arizona, where he graduated in 1996 with his master's in history. Leasure joined the Academy in 2001 as a history

29. The title alternates throughout the years as either Headmaster or Dean of Academics.

teacher. Leasure took it on himself to personally get to know every cadet and faculty member by name before the end of his first month. Surprisingly, he accomplished this difficult task within a week. While dedicated and determined, he never appeared stern; he always was in a good mood when greeting cadets. Leasure did not isolate himself in his office but rather was always out integrating himself into ANA life. Even when in his office, his doors were wide open for cadets. Leasure was a major element behind the progress of this era and left a legacy as one of the great Headmasters.

During this period, every academic department thrived with strong individuals. The friendly Ruben Rodriguez of the Foreign Language Department, Amy Coe of the Science Department and Sergeant Gregory Brown of the Leadership Department all knew how to befriend and educate unruly teenagers by partaking in cadet humor and getting to know inside jokes. The slightly stricter John Musser of the Fine Arts Department, Pamela Schneider of the Mathematics Department and Joe Monroe of the English Department knew how to gain the respect of cadets by not trying to talk down to them and instead treating them as equals. Instructors like Director of Studies Chair Sue Downey and Bruce Haggerty of the Social Science Department won cadets over by considering them as fellow seekers of knowledge, empowering cadets to absorb as much experience as possible. All these teachers had high levels of education and many years of teaching experience; all could have worked for any other schools for higher pay. However, for them it was not for the money; they wanted to stay because they truly enjoyed teaching. The small classroom size allowed for the building of strong student-teacher relationships. They could watch their cadets up close like an immediate family, becoming truly invested in their well-being. As chemistry teacher Amy Coe noted, "What I enjoy most about Army Navy is the close-knit family environment. My relationships with students don't end when the bell rings after class."

One of the Academy's more prominent faculty members from this period was Dr. Robin Cowen, who served as the biology teacher and faculty chair on the Board of Trustees. Cowen grew up in Winona, Minnesota, and got her PhD in plant breeding and cytogenetics from Iowa State University. While she lived most of her life in the Midwest, in the early 2000s she moved with her family to Southern California. She applied for the biology teaching position at the Army and Navy Academy. After not hearing back for a few weeks, she called the Academy about her application. She learned that the administration thought that she applied by mistake, as surely a doctor would not want to teach at such a small school. When Cowen informed them that she was serious about the teaching position, she was welcomed to the ANA family with open arms. Like the other teachers, she was an instant hit with cadets. She won the hearts of cadets by making learning a fun process. Anyone who grew up during my generation could instantly see a comparison between Dr. Cowen and Ms. Frizzle from *Magic School Bus*. Both were passionate, frizzled-hair teachers who wore science-themed dresses and made learning seem like magic.

LEADERSHIP EDUCATION AND MILITARY SCHOOL LIFE

The Senior Army Instructor is the next most influential figure on campus. He oversees the JROTC program. The role was first held by Major James Peyton, who served the San Diego Army and Navy Academy from 1927 to 1934 as the first JROTC Military Instructor. ANA is home to one of the top JROTC Units in the nation. Since 1971, the Warrior Battalion has won a gold star for Honor Unit with Distinction every year but one.

It is the Senior Army Instructor (SAI) who works the closest with the Cadet Corps. Two influential presidents, Colonel Cline and Lieutenant Colonel Miller, came to ANA as SAIs. During this period, the successor to these men was Green Beret Special Forces Major Michael Basista.

Major Basista continued in the footsteps of Miller by reviving old Academy traditions, and he took the Corps one step further by creating new customs. Basista wrote the new *Warrior Crest*, known to new cadets as the "Plebe Book," which ensured the continuation of Miller's practice that cadets learn the Academy's history. Weekly parade practice was perfected under Basista's guidance. With each successful parade, the major would give his trademark speech concluding with, "This was the best parade I have ever seen."

The Residential Life Department is the last important leg of the Academy foundation. Originally founded as the Home Department, operated by Bessie Davis, the Residential Life Department is now overseen by the Commandant of Cadets, a role that is the successor to the Superintendent. The Commandant of this era was U.S. Air Force colonel David Witwer. During this period, the Academy had a strong support team of TAC Officers who would act as on-campus guides to the boarding cadets. TAC, a term used in military schools, stands for Teacher-Advisor-Counselor or Tactical

Officer, depending on the school. While residential overseers have existed in some capacity in ANA's history, from dorm mothers to the housing staff, it was during this period that the role of the TAC Officer took on its familiar form. TAC Officers filled the role of the out-of-classroom teachers. While teachers in the classroom taught book smarts, TACs helped develop life skills and logical thinking.

One TAC who left a major mark on ANA was Jerry Thomason, who retired from the Marine Corps as a sergeant major and was the main face of Middle School residential life for Echo Company. If ever there was a reason to start the cadets in the Middle School, it was Thomason. While his loud "devil dog" bark and southern yawl might intimidate those in his presence, Thomason's jolly nature won him over with the young Middle School cadets, who looked up to him as a grandfatherly figure. His motivational speeches and study hall sessions helped instill strong discipline and improved academic grades of the young Echo cadets. The 2007 academic records, for example, showed that 99 percent of seventh-grade cadets scored a C or higher in their first semester grades, while 100 percent of eighth-grade cadets had a C or higher.

Under Thomason's influence, the Middle School Stoll's Guard was inspired to outperform the more experienced High School Davis Guard numerous times on the parade field. His role as Echo Company TAC Officer was so personal and sincere it would be impossible to emulate. The Middle School for most of the Academy's history appeared as just the obligatory extension. It was Thomason who shaped Middle School life into a valuable experience. Any school would be blessed to have a figure like Thomason at the helm.

Chapter 19

Brother Stands by Brother

In ancient traditions, it was always customary for my family to send sons off during their youth to train with military units of neighboring kingdoms before returning home to inherit leadership positions. Time passed, and my family moved to America. My grandfather served as an officer in the United States Army during World War II. My father served as a helicopter pilot and officer in the Marine Corps during the Vietnam War. Family tradition continued in a new context when my mother thought it would be best for me, as the eldest son, to be sent off to military school.

Not wanting to leave California, one day during a Boy Scout meeting I was thumbing through the pages of *Boy's Life* magazine. My eyes landed on an ad for the military academy in Carlsbad. At this point in my life, I had never left Northern California, much less the state, so a beachfront school in sunny San Diego seemed like an appealing choice for me. Before I knew it, I was on my first-ever airplane flight heading to Palomar Airport. I was given a tour by a cadet my age. I recall being embarrassed wearing a somewhat obnoxious wool suit my father gave me and eating my mess hall food in silence while the cadet guide tried to break the ice and start a conversation.

Hell Week

My head was freshly shaven. The sun was hot and beating down. I was standing statue stiff at attention in a formation of new cadets—"plebe" is what they called us. Senior cadets were barking in our faces, while hungry mosquitoes buzzed even louder. Every bite from the bloodthirsty bugs was followed by a growing intensity to fight the urge to scratch. Then we were down, all of us plebes doing push-ups. After another round of push-ups, my muscles ached. I could hear the waves of the Pacific crashing along the western border of the campus. I longed to jump into those waters to get away from the mosquitoes and senior officers. At that moment, the realization hit me—*what did I get myself into?*

During the first two weeks preceding the semester, all new cadets (plebes) had to go through a period of training called Hell Week. Modeled off boot camp, this initiation period taught cadets military procedure, drill, discipline, how to make beds and prep uniform and, most importantly, team bonding.

We went through countless hours of drill training day after day. Competitions were held, and senior cadets would bet on which plebe would come out the victor in drill downs. Occasionally, we would see old alumni watching from the sides, reminiscing on their time as cadets. In the evenings, we would remake our beds over and over until we could perfect the hospital corners. At night, we shined our two pairs of dress shoes and memorized all the information in the "Plebe Book."

Over those two weeks, I picked up a lot of ANA lessons. Mess is where we plebes ate our powdered eggs in silence. Ops meant Operations Office. Jarhead (a word normally used to describe a shaven Marine) referred to the disciplined cadets, often regarded as the brown-nosers of the class. Coach referred to John Maffucci. The ROTC brass torches worn on our lapel tips had to be handled carefully, as any fingerprint on the mirrored surface would result in a noticeable smudge. Names needed to be written in the cadet caps. My cadet number was 24014. Cadet officers needed to be saluted—if not, it would mean push-ups. That meant quick memorization of rank and a sharp ear looking at the shoulders of any approaching cadet in uniform. Company first sergeants were people to be feared yet not saluted, or that would result in more push-ups.

For all new cadets, it was mandatory to initially join a sports team, so I signed up for cross-country, a sport for which I was not cut out. I was overweight and started out running fourteen-minute miles. However, for three days I accidently joined the football team for training in the weight room. I remember being confused why cross-country runners would need to train their upper body. When I finally did join the other runners, I got lost on my first run, which resulted in the coaches searching for me in their van. When those two weeks of Hell Week finally ended, I had earned my cadet ranks. I had also brought my mile run time down from fourteen minutes to five.

Alpha Company

My first roommate and lifelong friend at the Academy was Ron Ruangtragool. He was a year younger than me but skipped a year, and we were both starting our freshman year at ANA. Ron enrolled the year before me as a Middle School cadet and helped me adjust to campus life.

At ANA, the dorms are divided up between various companies in a manner most cadets would compare to the Hogwarts houses from *Harry Potter*. There was the Headquarters (HQ), which housed the Battalion officers in Fegan Hall and the Davis Guard Drill Team and color guard in Crean Hall. Band Company was made up of permanent members of the Atkinson Warrior Band in Atkinson Hall. Alpha—considered the troublemakers, the company where all the rowdy cadets were sent to—occupied upper Anderson Hall. Charlie had lower Anderson. Bravo had the various cottages and the older Hoover and McClendon Halls. Finally, the Middle School Echo had McIntosh and Lewis Halls and any remaining rooms in Atkinson Hall.

The various companies compete in a year-long competition where everything becomes tallied for points—inspections, merits, GPA and more. This became a source of pride for cadets. The winning company earned a new streamer for the guidon and has a position of honor the next year in the center of the formation grounds and in front of the parade. Daily competitions include each company sending a cadet up to be inspected—the number of uniform gigs determines the company order for the mess hall line for that day.

HQ and Bravo, the most disciplined, were always the frontrunners in the competition. Ron and I were both cadets of Alpha Company, a company with a lot of pride but no expectation of winning the company competition. Ron was the Alpha guidon bearer. Ron and I were often sent up as the company representative for morning uniform inspections and monthly cadet competitions. Alpha was always last in the competition and hadn't won in years, and thus it was pushed to the very far corner of the formation grounds, literally just around the corner from Davis Hall, out of sight of Battalion Staff. Most of the company were habitually late (including the leaders) and put no care in uniform appearance, yet we had the most pride out of all the companies. There was a sense of identity, being a part of Alpha. We were the only company to have our own motto and signing-out call. When all the companies are dismissed by the adjutant, all Alpha cadets would shout "ALPHA!" do an about-face, bend down and swipe at their dress shoes with their left hands and then right (scuffing them) while simultaneously sticking their bottoms in the air directly at the commanding officers. Then they'd shoot back up, shouting a loud "YOU KNOW!" before departing their separate ways.

Alpha seemed to be the place where the latest basketball recruits from Chicago were placed, including my friend David Howard, no. 23 on the basketball team. Bravo had the more disciplined cadets. A lot of drill team recruits did come from Bravo. My friend, Arden Tran, started in Bravo. In Charlie, Ron introduced me to a friend from his previous year, Nathan Morales, who was known around campus for always taking pictures wherever he went for the Yearbook Club. I later realized that Nathan was the cadet guide who had showed me around my first day.

Living with Alpha in Anderson Hall was pretty noisy. Ron and I lived in the back room of the middle hall. We kept our area very clean; in comparison, venturing to any other Alpha area, you are instantly met with a heavy whiff of body odor. As Alpha was filled with rowdy cadets, it was where most of the craziness happens. In retrospect, I am surprised the Academy gave the top floor of the new building to Alpha. One Alpha cadet I vividly remember, known for always eating whatever he could get his hands on, was extremely obese and never took showers. When he was in the area, the smell was gag-inducing. After he spent two months not bathing, a large portion of Alpha grabbed him and got him into a shower. Bars of soap were thrown over in hopes of cleaning him—most of them had been purchased from an after-Halloween sale, so many were shaped like candy and sweet treats. After he left the shower, we all noticed large bite marks taken out of the soap.

THE GREAT TOWNIE FIGHT

During a cold, foggy night in November 2004, a Middle School cadet walked outside to head to the communal restroom. The fog outside was light, but the cadet could see in the shadows an enormous figure. At first, it appeared to be an older cadet, but then he noticed the long hair and civilian clothing. He saw another figure and then another. A gang of local high school students had wandered onto campus looking for trouble.

The cadet took off running, making a loud commotion as a nearby senior cadet took notice and quickly ran out to confront the outsiders. Another cadet pulled the fire alarm. All of Alpha woke up to the sounds of the fire alarm.

In the distance, I could hear, "Townies! Townies on campus!" referring to local high school students who went to Carlsbad and Oceanside High.

Seniors Echo were the first to approach the townies—they wanted to make sure the Middle School cadets were safe. The others quickly made their way to the Operations Office to get the TAC on duty. As the cadets approached, they noticed that some of the townies were swinging baseball bats.

"Townie fight!" I heard someone scream outside my door. Ron quickly darted out of his bunk to see the commotion. The flashing white lights of the fire alarms created a strobe effect that blanketed the entire campus. The night TAC called the police and then approached the townies, who were already swinging their fists. A townie ran swinging a skateboard towards a senior cadet, who punched the skateboard, breaking it in half with his bare fist. Another senior, a ripped body builder, tore off whatever clothing he was wearing to show his chiseled abs and muscles and then dove into the crowd, scaring away another portion of townies. Another cadet was able to wrestle a bat away from one man, who quickly scampered away.

By the time the police came, most of the townies were gone; the others quickly bolted at the sight of the police lights. The paramedics came and helped patch up the injured cadets. Luckily, aside from the senior's fist, all the other cadets escape any harm. This event became known as the Great Townie Fight of 2004, and it was a unifying event that brought the rival companies together.

Renovations

In the fall of 2004, the ANA Parents Association started a new series of renovations. The Academy at this point really needed new updating. The Recreation Hall and the dorm halls built in the 1940s were on the list to be renovated. It would mostly be done by parents and cadet volunteers.

One side project was digging through the old guardhouse. The building used to serve as the old Operations Office was where cadets would check in and out from campus. Later, the Operations Office was moved to its new location behind Davis Hall, and the 1980s guardhouse was converted into the Atkinson Museum. The museum housed the Academy's history, old uniforms, banners, yearbooks, photos and all sorts of trinkets. Notable items include a case containing the walking canes of the last Union and Confederate soldiers from the Civil War, boxing gloves signed by Floyd Patterson and the uniform of the Academy's old mascot, a dog named Ace.

The old building had no ventilation, so mildew and mold built up, destroying countless artifacts. After helping fix up the place, I helped run the museum for the next four years as an unofficial curator during parade weekends.

SKIP STEP

Nearly every month, the Corps would put on a dress parade. The most important of these parades occurred during Homecoming and Veterans Day, as well as Parents, Grandparents and Alumni Weekends. Thursday afternoon after class was the time specifically reserved for a parade practice. Uniforms would also change according to the day. On Monday, Tuesday and Friday, we cadets would wear our standard short-sleeve, West Point–style black wool uniform. On Wednesday, we would wear long-sleeve black wool with tie for Chapel, which took place before class. On Thursdays, for parade practice, we would wear Army Battle Dress Uniforms (BDUs) and, later, Army Combat Uniforms (ACUs).

Each parade began with the band in the front playing cadence, followed by the color guard, Battalion Staff and the previous year's winning company; finally, the remaining companies passed in alphabetical order. The adjutant would sound the Battalion through cycles of marches to get all the companies in place. The band plays "Sound Off," marching across the parade field playing all the armed forces marches. Then the color guard would come forth, and cadets were ordered to present arms. Then "The Star-Spangled Banner" plays. Then the Battalion Commander, the president and visiting guests would inspect the Corps. Finally, the adjutant would call the Battalion into different positions before marching in front of the viewing crowd and marching off the field.

This entire procedure is something most cadets dread; it must be learned by new cadets within a month. While for practice parades, the BDUs are bearable, for the actual dress parades cadets are required to wear the uncomfortable coatee uniforms. They are the same 1800s uniform worn at West Point, very thick with itchy Velcro wrapped around the neck, along with a toy soldier–looking helmet (shako). In the hot San Diego afternoon, standing still at attention sometimes for hours on end becomes unbearable.

We were told to slightly bend our knees to keep blood circulation flowing. In the back of the field, there are cadets with water on hand for anyone on the verge of fainting.

After a few hours of standing stiff in the sweltering heat, marching for the pass in review for the crowd does become a bit of a difficult task. The shako makes it extremely hard to see, so walking in a straight line does require cadets to keep an eye on the person in front of them. The one thing that is always observed by the crowd is the uniform in-step cadence of cadets in time with the drums. This is harder if the band decides to play a tune.

When a cadet is out of step, it is very obvious from a distance. For a cadet without the experience, it is very hard to get back into step. At ANA, there is a technique I learned; while there is no specific term for this, we cadets call it a "skip step," an extra skip on the same foot. The movement is seen as awkward but necessary to get back in unison with the rest of the unit.

After my first parade, the Senior Army Instructor, Major Basista, gathered all the tired cadets around and proclaimed, "In all my years at ANA, this is the best parade I have ever seen." I would learn by the next two parades that this is something he always said regardless, later saying that statement in a more self-aware manner.

I joined the Atkinson Band. While I was still a member of Alpha, I would march with the band for parades. My first band instructor was John Musser, a teacher who had been working at ANA in various position since the 1990s. I wanted to play the trumpet, but as I was wearing braces at that time Musser had me buy a flute. I was a flutist for the next two years—in my opinion a horrible one. But by my sophomore year I had become the woodwinds chair, marching as a squad leader in the parade and wearing a distinctive yellow plume on my shako.

As my first year ended, the senior class of 2005 gathered in the Senior Garden, singing the alma mater one last time as cadets in dress whites with white sunglasses stood nearby; Ron and I sat on the top steps of Fegan Hall looking down on the graduating cadets. We had just finished our freshman year and were discussing our next year. Ron would join the band, while I took his position as Alpha guidon. We were also the only cadets to skip sophomore Leadership Education Training (LET2) and go straight to junior LET3. As we looked down at the graduating seniors, we questioned why they wore those white sunglasses—we wouldn't know until we became seniors, and for us that was an eternity away.

Before They Do Unto You

First Sergeant Martin Davis was a memorable figure to cadets. His class was always a guaranteed source of amusement for those who got to watch him yell at cadets caught in his crossfire. During the average LET3 hour, the class leader, after we formed up, would knock on the door signaling to the first sergeant that class had begun. A large black man in army green, with a bald head and dark sunglasses, would then storm out. An intimidating figure, Davis walked to the front of the platoon, eyeing us like pests that had wandered into his garden.

Part of the class grade depended on daily uniform and hygiene inspection. Three gigs and we failed for the day. We all stood nervously as Davis walked up to his first victim. "How long you been wearing that shirt, boy?" Davis, a U.S. Army noncommissioned officer, carried himself in a different tone from the TACs, who were mainly Marine sergeants. When the cadet stated only "ten minutes," Davis jumped back with his eyes bugged out like he heard something profound. "Ten Minutes," he boomed as he walked right up to the cadet. "Your shirt is Nasty! You is Nasty! Your shoes are screamin' for Kiwi![30] Fail!" To the next cadet: "Young man! How long you been shaving? Negative days?" He pointed to the dormitories. "Now run to your room and Shave! You look like the Wolfman!" Inside the classroom, Davis yelled at a cadet with bad body odor. "You parasite! Look like you just crawled out of a damn petri dish!" He then took out his can of Lysol and started spraying the classroom, a common occurrence.

Davis was famous for his tests, giving us a binder of information to study. Before every exam, he would have us highlight topics he *guaranteed* would be on the test. LET3 cadets will remember these sessions. "Skip paragraph one, highlight all the words in paragraph 2. Now highlight paragraph 3 to 5.… Now highlight paragraph 6.…"

One day, a cadet asked, "Did you ever kill anyone?" Davis responded, "When we were parachuting into Granada, they started shootin' at us. Now shooting at parachuting soldiers, that's against the Geneva Convention, but they didn't care, and as I got closer to da ground I didn't care either. When I hit that ground, I didn't care if you were eight or eighty, crippled or crazy. And let's just say they didn't shoot back." He then finished with his catchphrase, "Remember do unto others before they do unto you."

30. Brand of shoe polish used by cadets.

Another cadet asked, "What would you do if someone ever broke into your house?" He looked at us and then stated, "I'd beat that [expletive] with a bat, then I'd tie him to a chair for a week and come back every day to beat him again. On the last day, I'd break his knees and let him go." Then, when we asked what he would do if the individual went to the police, he responded in a calm manner, "I'd put on my best uniform, march right into the police station, look that officer in the eye and say, 'Do I look like the kind of person who would do that?'" Funniest moment in LET3.

MIDDLE YEARS

With enrollments steadily increasing once more, this was the year Delta Company was reestablished. The company was housed in the cottages on the south side of campus. One cottage became infamous in the 1970s when the staff found that cadets had dug out a tunnel and an underground room under a foot locker. The underground room was said to be impressively large and housed a bunch of furniture and even some parade rifles from the armory.

By my junior year, I ended up becoming the Assistant S-2. There is a little tradition in Alpha Company of producing a long line of Battalion S-2 officers. The new S-2 chose me to be his NCO after I decided not to take the first sergeant role. Most cadets who aim to be Battalion Commander (BC) their senior year often try to get a first sergeant position junior year, although this wasn't something on my mind. Joining me in assistant staff was Ron (S-4) and Nathan (S-5). This meant our little group of friends got to room together in Crean Hall.

Our group of friends was often partaking in the same extracurricular clubs. Ron, Tran, Howard and I established the Interact Club on campus. We were also all involved with the track team, Beta Club, Key Club, Academic Decathlon and JROTC Competition. However, focus quickly shifted to SATs and preparing our college applications.

Around this time, I reformed the defunct Astronomy Club[31] and the Boy Scout Troop. For the latter, I started up conversations with the solitary Major Donovan, a former Marine who worked in the Ops Office for many years

31. I am happy to see that it is going strong, with a large group of interested cadets. It's also nice to know that the Astronomy Club is working directly with the Applied Physics Lab of my other alma mater, Johns Hopkins.

and once ran the Boy Scout troop in the 1980s.[32] He is well known to cadets for the last three decades as the default voice of the intercom announcements and the phrases, "Attention on campus!" and "Cadet please report to the Operations Office." We cadets considered his gruff announcements as part of the ANA experience. Recently, in an online alumni group, a cadet from the 2010s posted, "Attention on campus. It's raining outside. That is all." Alumni from as early as the 1980s immediately responded with their memories of the major.

Officer Candidate Course

I learned that my mother had been battling cancer. While my parents tried to keep this a secret, by my junior year my mother was already losing her hair. Eating was hard, and she had tubes strung throughout her body. Despite being in pain, she always convinced my father to drive the nine-hour journey to the Academy to see us every parade weekend. Around this time, I ended up focusing all my attention on academics, with the goal of attending medical school.

The end of junior year is the most difficult time for every cadet. The month after Spring Break, all juniors endure a month of the ANA Officer Candidate Course (OCC). During this time, mandatory SATs also take place for all juniors. Adding to that stress were final exams and, if one were an athlete, final sports meets and championships. We all heard stories about OCC from the previous class. Some seniors even made money selling their OCC notes to juniors, who were hoping to get a good officer position. I knew that I would not be able to relax for Spring Break and intended to use every moment studying. However, when my brothers and I arrived home, my father sat us down and informed us that our mother passed away during surgery just a few days earlier.

ANA staff individually sent their condolences throughout Spring Break. Dr. Cowen flew out to attend my mother's funeral. My brothers and I wore our Academy uniforms and laid our name tags on her coffin. Knowing what was to come upon my return, the Academy recommended I skip out on OCC and take some time off. However, when Sunday night came around, I was back with my classmates ready to face our long month.

---

32. I did end up reviving the Boy Scout troop, with the librarians as our advisors, and secured the number 1910 for our troop—fitting, as it as the founding year of both the Boy Scouts and the Academy.

We were all stripped of cadet rank and position and formed up as a separate unit from the Corps. During this period, we all had to take turns leading, studying for numerous leadership tests and preparing for the interview that would determine our placement for the following year.

Then came the highly anticipated and feared weeklong period at Camp Pendleton, where we were greeted right off the bus by hordes of barking Marines. We were huddled into a musky barracks, ordered to make our beds and cleaned the area countless times. The drill instructors greeted us every morning at 5:00 a.m. and had us crawl on our stomachs, scrubbing the bathroom floor with only toothbrushes. Then our days were filled with countless obstacle courses and challenges meant to put all the leadership skills we learned into practice.

Two track meets took place around this time. Juniors on the track team were shuttled from Pendleton to the competition and then rushed back. In one case, I had to go into a CS gas chamber with my eyes scorched from the burning flames, then was quickly shoved on the bus to run in the 1600m and 3200m races before quickly being brought back to repel off a five-story tower.

During the night, we got an hour of silence before lights out. This period was spent studying the enormous leadership packet we would be tested on at the conclusion of the week. Any extra time would be used to study for the SAT and exams, which took place the week after we returned to campus. One of those nights, we had an epic pillow fight that shredded the pillows down to scraps and feathers.

Tran and I ended up being tied for first place in OCC, scoring 99.8 percent and both awarded Honor Graduate plaques. Tran became the BC and I became the Battalion Executive Officer (XO, second in command). Along with Howard, the new sergeant major, we made up the new "top three."

At the end-of-the-year awards ceremony, I felt placed into an awkward situation, as I was called up to the stage again and again for different awards, including a Marine Scholarship, Army Scholar/Athlete award and a medal from the AMCSUS. Each time I went up, I would come back with a new plaque, medal or certificate. I was automatically singled out in front of the rest of the Corps and audience. By the time my name was called for a tenth time, a parent yelled out, "Just keep him up there already, for God's sake!" Other parents grumbled in agreement. The only thing on my mind was the fact that this would be the first of many ANA events my mother would not be attending.

Finally, the big moment came: the Change of Command Parade. It was a hot day; however, we did not notice the heat as much, as there was a cloud of excitement in the air. Tran as the incoming BC was standing with the graduating BC, separate from the rest of the parade. I took the long march from my position to the center of the parade field, where I saluted the graduating Battalion XO, relieving him of his post; he marched off to lead the large platoon of forming seniors. As I took his position at the center of the field, the announcer of the parade said:

> *Attention to orders! By the order of the President of the Army and Navy Academy: Whereas, the cadet leaders of the Class of 2008, having pledged to fulfil the obligations under the JROTC Leadership Creed and the Army and Navy Academy Leadership Oath. The Class of 2008 are hereby ordered to assume the command of the Warrior Battalion, with all rights and privileges thereto.*

The old BC relinquished his command by passing the Academy colors to the president of the Academy, General Bliss, who passed the colors to Tran, appointing him the new Battalion Commander. Tran held on to the flag for a moment and then raised it up over his head like a Spartan spear, sounding, "WARRIOR PRIDE!"

WATERMAN ACADEMY

Towards the end of the summer of 2007, my class returned two weeks early before the new cadets. Major Basista gave us lessons on how to effectively run the Corps. ANA partnered with a program called the Waterman Academy, a San Clemente–based group operated by members of the Coast Guard, for this purpose.

We boarded the ship the *Rapture* and sailed out towards Catalina Island. The first night, we had to do a swim test, and those who could not swim were given a life jacket to wear. I was somewhat embarrassed to admit I did not know how to swim. I jump out into the dark, cold water only to find myself quickly tossed a floatation device. The same happened to Howard. The next classmate to jump in was Russel Reed, a large individual on the football team who easily swam circles around us. The three of us had all started as freshmen in Alpha, where Reed gained a reputation of being a troublemaker. He always liked to joke around yet surprised us that night with his swimming abilities. However, just as the lifeguard was done rescuing any other flailing swimmers, Reed feigned swimming inability and was given a life vest. He would go on the next few days clowning around, making even the Waterman staff laugh.

The most memorable training event occurred on the second midnight. We were divided into teams for a mock rescue mission. To simulate a helicopter drop, we all had to jump off the top of the ship, five stories high, into the icy cold water. We were told the exact way we had to jump—there was no room for error or we could hurt ourselves. Our arms had to be wrapped around our body with our feet pointed down. This was mandatory for everyone. I was extremely nervous; I never dived off anything in my life, especially from this height. Thankfully, this took place at midnight; therefore, I could not see the distance of the water below me. I jumped and fell for what seemed like a minute. Before I realized it, my entire body plunged into the icy depth, sinking deep into the Pacific. My body stopped sinking and then held in place for a few moments. Then I started to slowly rise towards the surface. The ascent felt longer than the drop, as I was about thirty feet below the surface. Just as I exhausted the air in my lungs, I was grabbed out of the water and hauled onto the little raft with my team.

Each team member had to make that jump and make their way to the raft. Once on, we made our way to a distant cave to rescue a hostage. Two of us had paintball guns and protected the team from the instructors waiting in the cave with their own guns. As we approached the cave, two of us had to jump out and guide the raft through. It was difficult with the boat's motor in the way, sharp rocks below and heavy currents. We landed the raft on the beach and quickly rescued our hostage.

THE SENIOR PALACE

Battalion Staff, which included the top three (Tran, Howard and myself), Ron, Nathan and others, lived in Fegan Hall, the old Red Apple Inn, known to cadets as the "Senior Palace." The Senior Palace was run-down, being the oldest building on campus. However, it was a considerable distance from the other dorms, meaning we rarely had unexpected inspections. Because of the isolation, most of the seniors in the Palace would also get up to crazy antics.

One tradition for seniors in the Palace was a visit to the closed-off bathroom. There is one communal bathroom in the entire building, but when this place operated as an inn, each room had its own. One was boarded up behind a wall. No one would even suspect it exists aside from old lore told around the campus. There is a legend told that back during the days of the Red Apple Inn, a hotel guest killed a prostitute in the bathroom. The blood could not be cleaned, so the room was just boarded up and ignored. Seniors before us found a way into this room by going through the air ducts above. A tradition for all new seniors living in the Palace is to make their way to this room and sign their name on the wall. As there are no windows or lights, the dark bathroom lends itself well to such old ghost stories. It was also a place that was not accessible to anyone unless they were small enough to make the climb, so it was no surprise that by the time we got to explore this hidden room, it was filled with a sea of old cans, probably a collection that had been building up for decades from cadets long before us.

SENIOR RING CEREMONY

There are three major traditions that signify the ascent of a junior to a senior. One is the pinning on of officer ranks at the conclusion of OCC. The second is the bestowment of West Point sabers during the first parade of the year. The final is receiving the class ring during the Senior Ring Ceremony. While the ring tradition extends back to Pacific Beach, the ceremony itself was started by Virginia Atkinson in 1946. A new ring was designed to reflect the school colors—a gold band with a black onyx jewel surrounded by the Academy's name. Mrs. Atkinson's idea was to have the ring resemble those of West Point and the Naval Academy, and a new ring was designed exclusively for ANA, with the school crest on one side and an eagle wielding crossed rifles and anchors on the other side. As long as we were cadets, tradition had it that the crest would be turned facing us to remind us of our goal of graduating. Upon graduation, the crest would be turned to the world, showing our accomplishment. The ring is the administration's official recognition of the graduation class. The ring design remains the same six decades later.

Mrs. Atkinson proposed having a ceremony presenting the rings to seniors during the Autumn Ball, held during the first parade weekend of the year. Back then, the rings were first presented by Colonel Atkinson, who would ascend a stairway to the stage in the middle of the ball. The

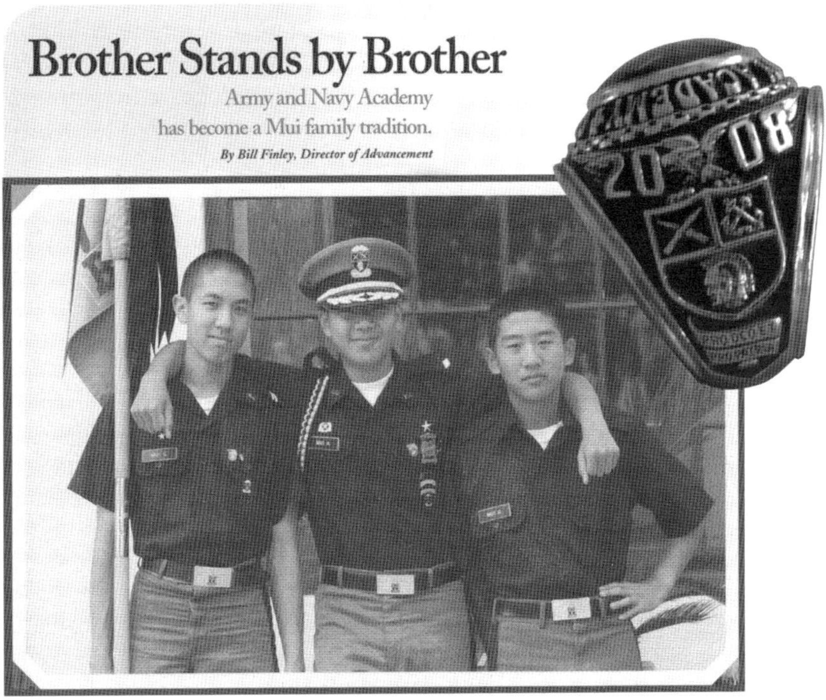

Article in *Army Navy* magazine featuring the Mui family, standing in front of Davis Hall—Bravo Company guidon Christopher Mui (Class of 2010), Battalion Executive Officer and Honor Council president Alexander Mui (Class of 2008) and new plebe Gregory Mui (Class of 2011). The class ring belongs to Alexander. *Author's collection.*

president of the class would lead a long line of seniors to the top of the stage to receive this long-awaited symbol of the first class.

Later, the tradition was changed when the task of presenting the rings fell on the ladies who the seniors took to the ball. The dates would wear the class rings on heavy ribbons around their necks. When they walked onto the stage before a large replica of the ring, they would place the ribbons on the necks of the seniors as photographers took pictures. By the time I received my ring, the tradition occurred as its own separate event before the Autumn Ball in the Virginia Atkinson Chapel with family in attendance, taking place before a large replica of the class ring.

Poinsettia Fire

On October 20, strong Santa Ana winds and severe drought in California caused massive wildfires to break out. The roaring fires eventually covered an area of 1,087,110 acres, visible from space. Governor Schwarzenegger declared a state of emergency, and local public schools started to evacuate. In Carlsbad, the Poinsettia Fire, considered the most destructive of the San Diego County fires, burned several hundred acres, causing private schools in the area to evacuate their campuses one by one until ANA was the only local school still in operation.

For five days straight, ash rained down like a heavy blizzard, coating the entire campus in a thick layer of white-gray powder. We could smell the smoke thick in the air and feel the heat of the fires, which were just miles away. The administration held off evacuating because it did not calculate any emergency dates into this academic year. Normally, five days would be put aside in case of events such as this, but as this was the first year they wanted to aim for a May graduation, there were no extra days to spare—all days missed could not be made up. Finally, days after the last school in our area had evacuated, I was called up to the President's Office. By now, I had learned that the fires were just about a mile from campus and that it was now urgent the school be evacuated. As the acting BC for the week, I organized the cadets for a mandatory evacuation. All cadets were to call their parents and quickly arrange a trip home within the next twelve hours. Of course, there was a large population of international students who needed to be sent to the homes of neighboring cadets and faculty. The campus quickly emptied as the fires loomed in the distance.

Let's See How Far We've Come

When we returned, a layer of ash still covered the campus, although within a day it was quickly cleaned up. At the Veterans Day Parade, Tran, Howard, Ron and I were awarded the Atkinson Silver A for our dedication to the ANA community.

As the first semester was drawing to a close, seniors had to help prepare the Corps for the Annual Federal Inspection (AFI), an event during which all JROTC units and military schools are inspected by the Cadet Command. AFI was always a source of pride for the school. Not content to merely pass, every year there was a push to earn the highest score, Honor Unit with Distinction (96 percent and higher). Since the JROTC established the AFI back in 1971, ANA had won the title Honor Unit with Distinction every year except 1989. That class was always used by Basista as a reminder: "You don't want to be like those '89 guys, do you?" Every time 1989 alumni are on campus, cadets point out that fact as their only notable "accomplishment," to which they would reply with a sigh, "Yes, we are *that* class."

Around this time, my class had developed a reputation for being rowdier than the previous years. Most of our teachers accepted us as just "boys having fun," while some did not know what to make of it. Despite the unruliness of our class, we were able to manage the Corps more effectively that our predecessors. We all saw one another like we were a family and, when needed, pulled together to get things done. Despite initial worries, we pulled off one of the best inspections seen in some time, with a score of 99.4 percent, easily earning the Academy its thirty-seventh Honor Unit with Distinction title.

The next month, Carlsbad High School celebrated its fiftieth anniversary. When Carlsbad High opened back in 1958, a parade of cars and buses took students, who had previously been going to Oceanside High, to the new school. And waiting along Carlsbad Boulevard, ready to welcome the first class of Carlsbad High as they crossed the Oceanside-Carlsbad border, was a formation of ANA cadets, all in full dress uniforms at attention with the Warrior Band playing "Welcome Home." The same bus trip was re-created fifty years later on Saturday. Wanting to honor the event, I decided to arrange an unofficial tribute by gathering the remaining cadets on campus in full dress uniform to salute, much to the surprise of the passing motorcade.

The month following was the weekend of our Military Ball, with all of us seniors and juniors performing the Grand March in Davis Hall. It was an impressive sight for all spectators, given the size of the combined classes. We lined up in promenade order and marched with our dates up and down the court, linking arms and crisscrossing in precise maneuvers all to the tune of Philip Egner's "West Point March."

This was my last year of the track team. Tran, Reed and I were team captains, along with our classmate Marcus Wilson. During the last meet of the year, Wilson was asked to sub last minute as the last runner of the final 4x400m relay race. Wilson had just come off another race but jumped right in. As the race started, our team gradually fell behind. By the time the third runner passed the baton to Wilson, our team was about half a lap behind the other runners. The moment Wilson grabbed the baton, he made a mad dash, nearly catching up to the other racers. Towards the final round to the finish line, Wilson had just about caught up with the other runners. Even the other teams and coaches cheered him on. He was then neck and neck with the last two racers, just about to reach the finish line, when his entire body convulsed and locked up, dropping him just feet from the finish line. Although he didn't finish the race, everyone who was there could see that he gave the most impressive performance of the event.

Many classmates, including Howard, Sherif Attia and Orlan Jackman (who would later play for the London Lions and Worcester Wolves), were on the basketball team, which won its first Conference Championship in two decades and became semifinalist in the Pacific League. The next year's team would go on to become Coastal League Champions.

The Warrior Hockey team had formed two years before. Despite being forced to travel two hours to Corona for every practice, the men dominated in regional competitions and won the San Diego Bronze Division championship those two years. To help support the new championship team, $30,000 was given by the ANAPA to help fund the building of on-campus hockey facilities. The ribbon cutting was held on April 4, 2008, as the new rink was opened.

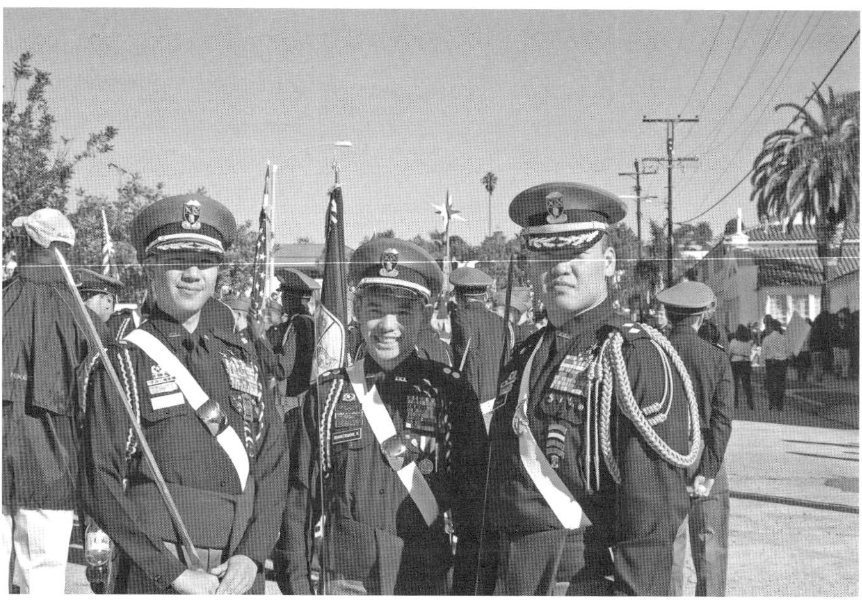

Cadet Executive Officer Mui, Operations Officer Ruangtragool and Battalion Commander Tran about to lead the Warrior Battalion and Atkinson Warrior Band at the 2007 La Jolla Christmas Parade. *Author's photo.*

Cadet's Final Command

We woke up the morning of May 31 for a breakfast on the Senior Lawn. Afterward, an award ceremony was held where I was once again crushed under a weight of plaques and awards. This time, the same parents who had complained about my name being called numerous times the previous year were now cheering me on with my family.

I stood at the center of Maffucci Field, slightly shaking with the excitement the day had brought. As my racing mind was trying to process the events to come, a voice came over the press box:

> *Good afternoon ladies and gentlemen! The Army and Navy Academy Corps of Cadets would like to welcome you to our Change of Command Parade. Today's review is in honor of our 97th Graduating Class, the Class of 2008….Cadets at the Academy have honored the traditions of our school through the years by gathering on the parade field and marching together as a unit….Today the Corps will parade for your review, led by the 97th Graduating Class. The Commander of Troops, Major Alexander Mui, formally presents the Corps.*

That was my cue. I called the Corps to attention, launching the start of the Change of Command Parade with, "Bring your units to attention!" During the next hour, I commanded the Corps through Sound Off, Inspection of the Troops, the national anthem, the Arch of Sabers, the Eighth Grade Bridging Ceremony and the Company of Year winner announcement. After all the events were completed, the announcer finally said, "And now, the final ritual of this afternoon's review, the official 'Change of Command' and the pass in review to honor the Class of 2008….The Battalion Executive Officer now gives the order for the current commanders and leaders to post their relief."

In what would be the last command I would give as a cadet officer, I sounded at the top of my lungs, "Post the Relief!" As the juniors moved up, the seniors started to form a platoon near the reviewing stand. I saluted my successor as I joined them. Then the announcer said, "Attention to orders! By order of the President of the Army and Navy Academy: Whereas the Class of 2008, having fulfilled the tenure of command as staff of the Warrior Battalion, the Senior Class is hereby relieved of duty."

Tran passed the school colors to the new BC and quickly joined us. The new cadet leaders called the Corps to attention and marched them past us.

> *Today's review marks the final formation of the Class of 2008, as cadets. This occasion is celebrated with a final pass in review and gathering of the graduating class at the reviewing stand to be honored by the Corps. Ladies and gentlemen, please stand for the graduating class! I present to you the Class of 2008!*

We marched off the field to cheers as we made our way across the street for the final event of the day. We lined along the outside of the Senior Garden and then walked through the archway with our class advisor, art teacher Terry Matsuoka, as the Warrior Band played "Pomp and Circumstance." We took our seats in front of Davis Hall as the flag was lowered from the flagpole and final awards were given. Finally came the time to give my salutatorian speech, which ran longer than anyone expected—perhaps this was the reason why the following year they put a time limit on graduation speeches. I looked at my seated classmates, all of us in white uniforms, donned with every ribbon and medal we ever earned as cadets. Everyone had their white sunglasses on. After Ron gave his valedictorian speech, one by one we went up to receive our diplomas.

We walked through a procession of faculty, who congratulated us as we gathered into the Senior Garden. For the last time as cadets, we sung the alma mater as tears of welling emotions rolled down our cheeks, obscured by our white sunglasses. As we sung the last verse, we all tossed our covers up into the air and gave a loud cheer, "08!"

PART VII

TOWARDS THE CENTENNIAL

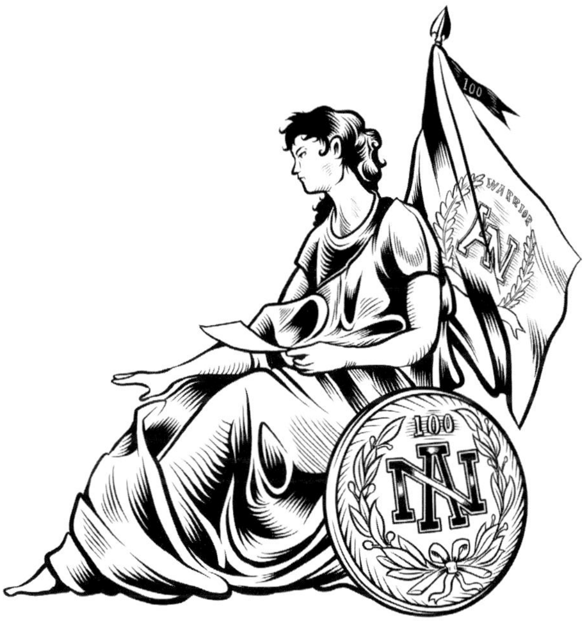

The Academy's Centennial. Artwork by Alexander Mui.

CHAPTER 20

AN INDUSTRIAL AGE

EVERYTHING CHANGES. The once wide-open campus now had cut itself off from the outside world with a large black spiked fence. The eight leadership traits embedded into the south front of Davis Hall had been reduced to six. The Rudyard Kipling poem under which my classmates formed every morning was replaced by a colored vertical advert. My former cadet family, Alpha Company, had been disbanded, and the guidon that I once wielded was stuffed back into storage. Within months, the place that I had called home for four years looked different.

The most alien artifact was what was in the place of the old flagpole. Standing before Davis Hall was an inapt yacht yardarm. The odd relic was structured like a cross with three additional halyards on each arm and an extra one in the front. Originally, the American flag was hanging on the front halyard, in accordance with maritime regulations for displaying a national ensign when out in sea. However, the passing public, not familiar with the guidelines for nautical flagpoles, was shocked to see the American flag hanging from the side under the Academy colors. Some alumni also pointed out that on land, the national flag needed to be raised higher than all other flags, so the flags were quickly switched. The two inside halyards flew four signal flags each, spelling out "ARMY" and "NAVY." One line flies the flag for the highest-ranked officer on campus, one flies a flag bearing a gold *A* when a "Golden A" alumnus visits campus,[33] one flies the

33. Like the Silver A, which is the highest honor an ANA cadet can earn, the Golden A is the highest honor an alumnus can earn.

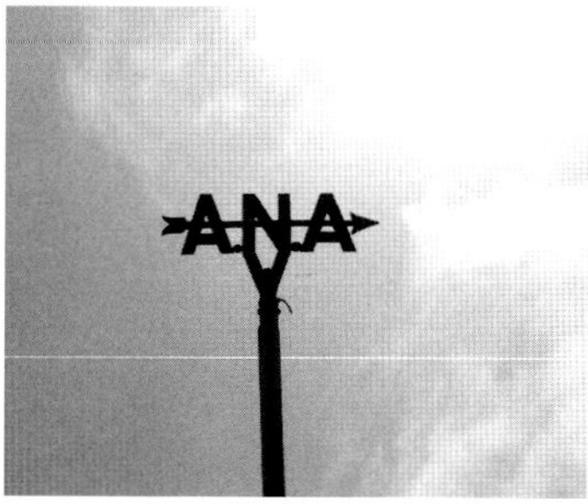

The Truck, given by Major General Joseph Fegan, Commanding General of Camp Pendleton, as a symbol of friendship between the Army and Navy Academy and the Marine Corps Base in 1948. The arrow points in the direction of Camp Pendleton. It was removed sixty years later, along with the old flagpole, for a new yardarm. *Author's photo.*

Centennial flag and another flies the Warrior "battle flag" denoting which sports event is taking place on campus—black represents home games and white represents away.

STOCKDALE FAMILY

The Stockdale family became involved with the Academy the school year after I graduated. By this time, the Academy's vice-president for institutional advancement was Stanford Stockdale, the son of famed Vice Admiral James Bond Stockdale. Admiral Stockdale is one of the most decorated officers in the history of the United States Navy. He had been awarded the Medal of Honor, three Navy Distinguished Service medals and countless other medals. He was the highest-ranking naval prisoner of war during Vietnam, where he was detained for seven years. In 1992, he was the vice presidential candidate on the Ross Perot Independent ticket.

The Arleigh Burke–class guided missile destroyer was named in his honor, the USS *Stockdale* (DDG-106), which was christened on May 10, 2008, in Coronado, California. During the fall of that year, the Army and Navy Academy formed a relationship with the crew of the USS *Stockdale*, with the varsity football team wearing the *Stockdale* patch on their uniforms. The crew of the *Stockdale* also invited the Cadet Corps for training on the ship during the summer months and would attend at home games when docked.

The special September 11 home game in 2009 saw more than thirty sailors from the USS *Stockdale* in attendance as they cheered the Warriors to a 30–13 victory against the San Pasqual Academy Dragons. During halftime, the commander of the ship, Fred Kacher, who served in the Pentagon during the September 11 attacks, was presented a framed Academy jersey as a token of appreciation for this new friendship. Commander Kacher thanked the Academy, saying to the crowd, "We are thrilled to hang the framed football jersey in our crew's mess, and I am thankful that the Army and Navy Academy honors Admiral Stockdale's legacy as we remember the fallen and their families on the eighth anniversary of the attacks on our nation."

New Changes, Old Alumni

After our first semester of college, Ron and I returned to campus to visit our old ANA family. We were both abruptly greeted with an imposing fence, an odd flagpole, a faded billboard pasted on Davis Hall and news that our home company, Alpha, was disbanded. We went to visit one of our old teachers and came across the recent issue of *Army Navy* magazine. The cover stood out from the previous issues of the magazine, which always prominently featured cadets. This cover had two bald men with glasses, dressed in suits, standing together with strangely creepy smiles, with the caption, "Helping move the Academy from Good to Great." These were the new Headmasters.

The Dean during my time, the innovative Derrick Leasure, was highly respected among both the faculty and cadets. He left as my class graduated for another job across the country. The community knew that his successor would have big shoes to fill. Among us it was unanimously assumed that his Assistant Dean, Robert Norton, would fill the role. Norton was the perfect choice; as Assistant Dean he was already learning the position from Leasure, and being a former navy officer, he fit right into the military school environment. Norton had already engrained himself into the community from his teaching positions, coaching duties with various sports teams, plus he had family enrolled.

It was no surprise that Ron and I were puzzled when we learned that Norton was not chosen as Leasure's successor and that he and the rest of the faculty had been passed over for a newcomer. Our former teacher also echoed our confusion, pointing out that an already established faculty member should have been promoted to the vacant position and that the new Headmaster had already gotten off on the wrong foot with staff and cadets.

However, it was the other figure on the cover who baffled us more. The new Headmaster of the Lower School was someone who was hired from the existing staff, although someone my classmates and teachers would agree was the last person they would ever choose for that role. ANA is a close-knit family. All cadets know the teachers and staff, and vice versa. Even if a cadet never had a class with a particular teacher, they were so involved in cadet life from organizing field trips and clubs, cheering at sporting events that it would have been impossible to avoid them. However, there was one who was the complete opposite, always lurking in the shadows. He was a figure I personally can't recall. Unlike other staff, he was anti-social and rude and made a great effort to avoid everyone. He was the last person any of us assumed would be the Lower School Headmaster, with one of our teachers commenting, "I have no idea what this guy had to do to weasel his way up to that position." This was a similar view echoed by the other staff and classmates.

During my time, this person was the de facto head of Middle School, although cadets didn't see it that way. For us, Sergeant Major Jerry Thomason was the undisputed face of the Middle School, motivating younger cadets to even outperform the High School in drill and academic standing. His nightly study hall sessions pushed Echo to have the highest academic standings. If there was a new head of Lower School, Thomason was the undisputed choice.

However, we learned that not only was Thomason not promoted, but he was actually let go to make room to hire new employees with "college educated qualifications." This shocked Ron and other former Echo cadets, who saw the old Marine as the staple of Middle School. Other staff and alumni also agreed, believing the administration to be out of touch with the ANA family. Thomason's knack for motivating young cadets was so honest and sincere it was something no college degree could come close to substituting.

Afterward, I wondered if my classmates and I were just being reactionary, as we were just young and naïve alumni, looking at the school through a nostalgic lens. However, as time progressed, it became exceedingly clear that the insights of the faculty and intuition of cadets were indeed correct.

The new Upper School Headmaster could not adjust to the new environment, and the cadets quickly rebelled against him, one major factor being that he turned a blind eye to cadets being bullied. Within a few months, he was quickly demoted, and longtime teacher Denis Anderson once again became Headmaster, as he had been back in 2001. The community agreed that he was the perfect choice.

The individual who became the Lower School Headmaster later made regional news when he was arrested for various crimes. While my classmates and teachers had no knowledge of his true character, we instinctively didn't like him for some reason we could not place. Our intuitive judge of character was finally vindicated upon his arrest.

THE REBRANDING

From the perspective of us young alumni, the administration was trying to combat the recession by trying to mimic other successful preparatory schools to make ANA more appealing to perspective families. It appeared that it was trying hard to rebrand and make the school more marketable, with new headmasters, new staff, new logo, new mission statement, new fence, new flag and flagpole, new front of Davis Hall, new building plans, new everything. Most of these changes were arbitrary, like renaming the TAC Officers as Residential Life Officers (RLOs) or changing the position of Academic Dean back to School Headmaster. Other changes actually appeared to make things more difficult and confusing. For example, Alpha Company was disbanded, so the companies could then be organized from Bravo to Echo. Although the elimination of a full company was done to combat sinking enrollment brought on by the recession, getting rid of Alpha Company defeated the alphabetical organizational purpose of such names.

In another change even more perplexing, the former High School (ninth through twelfth grades) and Middle School (seventh through eighth grades) were reorganized as the Upper School (tenth through twelfth grades) and Lower School (seventh through ninth grades). This created even more unnecessary issues, as the JROTC program is only for students in ninth through twelfth grades, who would need to be under different insurance plans. It made perfect sense to keep the cadets divided between those who were in the JROTC and those who were not. All the younger cadets not in JROTC were all previously part of Echo Company and had their own separate part of campus and dorms, away from the older cadets. Now the new Lower School had these Echo Company cadets plus the weird auxiliary of ninth-grade JROTC students who were scattered throughout the various companies, making simple things, like separate JROTC courses and formations, even more unreasonably difficult. This made the Bridging Ceremony that welcomed graduating eighth-grade cadets to the High School meaningless.

By now, we were beginning to feel like the older alumni I remember from past Alumni Weekends. A common theme with alumni, from fifty years ago or who just graduated the year before, were the opinions, "It was better in my day," or "My year was the last true ANA class before these changes." Perhaps this is the rite of passage for all Academy alumni, as my classmates and I were echoing the exact same phrases. However, for the Class of 2008, our year saw the departure of influential faculty like Derrick Leasure; teacher

Dan Loden, who had been with ANA for nearly forty years; and a handful of other faculty who departed at the conclusion of the 2007–8 academic year, all of whom we considered "honorary graduates" of our class. For us, the change was visible with the new Centennial logo and flagpole. If there was a class to state that things have changed since they graduated without being hyperbolic, it would definitely be the Class of 2008.

Chapter 21
Centennial Celebrations

The year was 2010. The United States was recovering from the recession. On October 10, across the globe, the Republic of China on Taiwan was celebrating the 99th anniversary of the Xinhai Revolution and subsequent founding. At home, Lauren Faust's *Friendship Is Magic* aired on television, marking the start of the Animation Revival. The day of 10/10/10 was indeed a day that represented renewal, celebration and special significance around the world. This was also the weekend the Army and Navy Academy chose to celebrate her 100th anniversary.

Years of planning finally coalesced into this one weekend. Across the region, newspapers and media groups all ran articles and promotional pieces commemorating the event. The initial outlook surrounding the Centennial Celebration eerily echoed the events surrounding the twenty-fifth anniversary back in Pacific Beach. In both cases, the Academy had just experienced a drastic drop in enrollment numbers brought on by major economic disasters that crippled the nation and yet she announced major building and expansion plans alongside heavily promoted anniversary celebrations to keep spirits high and hopeful.

Memorial to the Academy

The first event to open the Centennial Weekend was the dedication of the new Army and Navy Academy Veterans Memorial. The memorial was the

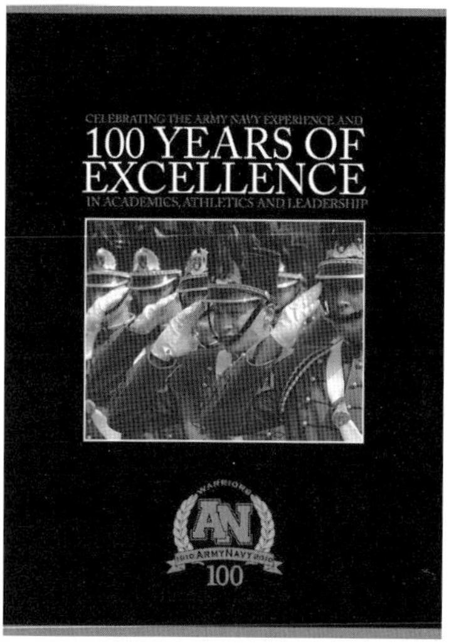

A promotional insert in the *San Diego Business Journal* covering the Academy's Centennial. *Author's collection.*

brainchild of two alumni from the Class of 1957, Buck Buchanan and Jim Christie, who noticed that there was no memorial to honor fallen alumni anywhere on campus. A committee was formed, and members of the Classes of 1955, 1956, 1957 and 1958 donated to see the memorial made a reality.

On Friday morning of the Centennial Weekend, the Cadet Corps unveiled the new memorial. Jim Christian, Class of 1956, gave the dedication address: "We gather here today to express our sorrow, and more importantly, our gratitude to our fellow cadets, who, over many decades, have sacrificed their lives for their country and for our freedom....Today we honor them."

Jack Jackson, Class of 1955, unveiled the memorial, and the Corps saluted. The memorial was a humble, inconspicuous plaque embedded into the cement at the base of the new flagpole. Made of solid black granite, these words were engraved on it:

> *This Memorial Is Dedicated to All Cadets Who Sacrificed Their Lives While Serving in the Armed Forces of Their Country.*

The words were chosen with deliberate care. The memorial is unique in that ANA was home to countless international cadets, yet it honors not just alumni who served in the Armed Forces of the United States but also all alumni who made the ultimate sacrifice in service of their homelands.

Another memorial was rededicated during the Centennial Weekend. The Class of 2010, which graduated in the spring of that year, donated to ANA a large black granite obelisk. Standing in the north patch of grass outside Davis Hall, the four-ton obelisk bears a brass relief of the Academy's mascot and honors "One Hundred Years of Excellence in Academics and Leadership."

More than one thousand alumni and fans filled the stands for the Friday night Centennial football game between the Army and Navy Academy Warriors and the Escondido Charter team. The crew of the USS *Stockdale* provided the color guard and sang the national anthem before the start of the games. The Warrior Pride Pep Band played music throughout the night while the crowds cheered on the home team. Luckily, unlike with the fiftieth-anniversary game, the Academy was able to play on its home turf. Unfortunately, like the homecoming game fifty years before, the Warriors ended the evening with a 14–48 loss.

Saturday was the day of the much-awaited Centennial Gala at the La Costa Resort. The Academy promoted the event as the "Party of the Century." As a Taoist, I was surprised to be approached an hour before the ceremony to give the opening invocation, which I gave in tribute to cadets of ANA's one-hundred-year-old "Long Gray Line."

More than five hundred alumni and members of the community were in attendance, including the last surviving graduate of the Pacific Beach campus, Jim Atkinson (Class of 1952, son of Colonel Atty), Richard Ede (Class of 1964, son of Headmaster Ede), Karen Miller (wife of the late Colonel Miller), Carlsbad mayor Claude "Bud" Lewis, Jack Wyatt (Class of 1963, Board chairman) and Colonel Donald Cline, who was visiting the campus for the first time in nearly twenty years. Another honored guest was the beloved mess hall chef, Joe Orozco, whom all the alumni clamored around recalling fond memories. Joe remarked to me that he felt out of place relaxing and eating food, as he was used to being the one who served the meals at events such as this over the last forty years.

Alumni who remembered Colonel Atkinson were greeted with a surprise when the voice of the old colonel was heard over the microphone opening the gala. Found in the time capsule buried in 1960 was a recording of Colonel Atty intended to be played for the Centennial.[34]

The time capsule, which was considered lost, was finally recovered just months before the Centennial in the trophy cases in Davis Hall. Inside was a letter written by Oscar Fitzgerald, the president of the Class of 1961, for the president of the Class of 2010, Charles Morone. Fitzgerald was at the gala to present his letter to Morone, which was read out for the attendees.

The new logo and flag were proudly presented to welcome ANA into the next century, and trumpets were sounded by bandmaster Major Mike Donovan and Lower School Commandant Major Charles Glimka. Playing on the screen was a historical montage, put together by alumni Joe Burden (Class of 1963) and Ed Roe (Class of 1960), that began with a tribute video I produced, made up of historic photographs from the Academy's past.

34. Originally, it was announced that the audio recording was too old to be played, so hearing Atkinson's voice was a welcome surprise for all in attendance.

The Centennial dress parade was held the next day. "The weekend was a total integration of the school wide community. Patrons, past patrons, alumni, faculty, staff and cadets joined together in the weekend-long tribute to 100 years of academic and leadership excellence," commented Bliss on the success of the weekend. "With the first century behind us, the Academy now turns its attention to the next century. We accept the challenge to continue to improve the Academy and ensure its future, and relish the opportunity to do so."

NEW EXPANSION

Both the Army and Navy Academy's Centennial Celebration and twenty-fifth anniversary came on the heels of major economic collapses. In both cases, the Academy bolstered support by using these events to announce major building and expansion plans. In the case of the 1935 event, Colonel Davis used abrupt announcements of his expansion to keep up spirits and take the focus off the recent bank takeover and financial problems facing the Academy. However, the 2010–5 Strategic Plan that was announced during the Centennial had been years in the making.

The Master Plan had been in the works since the early 1990s and had evolved many times over. One version included demolishing the 1940s dorms and extending Anderson Hall, another had a massive library sitting in the place of the old Red Apple Inn and one even proposed an underground tunnel from under the old guardhouse to Maffucci Field. During my time as a cadet, I recall seeing a small-scale model of expansion plans for the campus in the front trophy room of Davis Hall. This model occasionally threw visitors off, as it had buildings on it that didn't exist. That proposed design was scrapped long before I enrolled.

The current Strategic Plan was initiated in December 2009 and focused on making a new Master Place a reality. This was followed by six months of in-depth meetings, focus groups and involvement from all sectors of the community. This Master Plan was actually eight years in the making, starting back in 2001 under Miller. The plan saw numerous revisions over the years and was finally approved unanimously by the Carlsbad City Council in January 2010. In June, the Strategic Planning Conference was held to put everything into motion.

The twenty-phase Master Plan includes replacing many old buildings across the sixteen-acre campus, including new dorms flanking Anderson

Hall, a new Admissions and Administration Building and a new building that will contain a library and sky bridge connecting the west campus to the athletic field over the old Red Apple Inn stands.[35] This is meant to increase indoor space from 157,000 to 288,000 square feet and living capacity from 320 to 440 boarding occupants. It is predicted to cost $60 million.

Celebration Continues

The month following the Centennial Celebration, ANA co-hosted, with the Carlsbad Rotary Club, the Second Annual Veterans Day Memorial Parade and Wreath Presentation for the local community. This annual event is the brainchild of Colonel Russ Hanthorn, president of the local Rotary Club, who approached Bliss with the idea in 2009. This new event honors the veterans of San Diego County with a full dress parade and a ten-man color guard bearing the flags of all five branches of the United States military. The oldest veteran in attendance is designated the Honorary Commander of Troops. In recent years, the parade has ended with a special flyover by the Red Stars Formation Team doing the poignant "Missing Man" formation. The celebration concludes with a dedication to the veterans and a laying of a wreath at the Vietnam Veterans Memorial on campus. This new celebration is now considered one of the top events in all of San Diego County that honor veterans. Senior Army Instructor Major Basista noted, "This Military Review and Remembrance has grown from an Academy ceremony to a premier event. Our San Diego County veterans deserve no less."

During the planning of the Centennial, the Academy received a surprising honor, as the Atkinson Warrior Band was invited to perform at the National Memorial Day Parade in Washington, D.C., the following spring. Delighted, the band prepared the entire year for the event. The old Atkinson Warrior Band has seen revitalized success under the guidance of Major Michael Donovan, who had previously served as bandmaster at New Mexico Military Institute for more than two decades. Major Donovan was brought on as the new bandmaster in 2007, and during

35. Curiously, the new Master Plan does call for the demolition of Fegan Hall (old Red Apple Inn), which has been marked by Carlsbad as a historical site. However, the city has approved this plan, perhaps since the building replacing the old hotel will include a walkway over Carlsbad Boulevard that will state "Welcome to Carlsbad" on the north side, visible to drivers entering the city from Oceanside, an appealing feature to travelers.

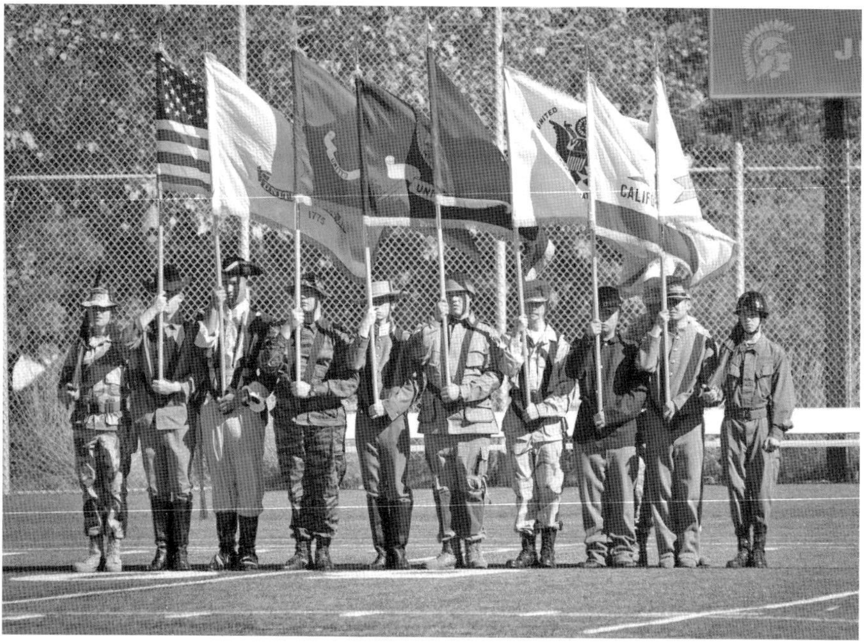

The Warrior Battalion Color Guard with the national flag, flags of the military branches, state flag and the new Centennial flag, with cadets in the uniforms honoring the soldiers who fought in wars of various eras. *Courtesy of ANA's Office of Advancement.*

this time, he brought the band to heights not seen since the time Colonel Atkinson served as band director.

In May 2011, the Warrior Band flew across the country to the nation's capital, playing at the World War II Memorial the day before the parade. The next day, drum major Lamar Greenwood led the Atkinson Warrior Band in the National Memorial Day Parade as they were cheered on by spectators, including a large number of alumni who turned up to show support.

After working four years with the Army and Navy Academy, Major Donovan retired on what can be said to be a high note—no pun intended. Recounting the Memorial Day Parade, he said, "It was the best band trip I've ever done." He had previously done three trips to Washington, D.C., when he was bandmaster of the New Mexico Military Institute, yet he was proud to state, "This is the finest group of students I have worked with in over thirty-five years of teaching, and it was truly the trip to end all trips!"

My younger brother Christopher Mui is Class of 2010. My youngest brother, Gregory Mui, is Class of 2011. Both of their classes laid claim to the title of being the "Centennial Class."[36] Being the year that welcomed the Centennial, the Class of 2010 donated the Centennial Obelisk to honor the Academy's past. Being the class that hosted the Centennial Celebration, the Class of 2011 wanted to donate a gift that symbolized looking ahead. As their gift, they installed a large compass rose in the sidewalk at the front entrance of campus. The compass bears the words "Never Lose Your Direction" and is meant to point the way forwards to all following cadets. In a sense, this new landmark is also meant for the Academy herself, pointing the way for a new century and bright future.

36. The title Centennial Class was chosen when it was revealed that neither class was the 100th graduating class.

Chapter 22
Into a New Century

On October 4, 2013, during halftime, the outgoing president was called on to the newly christened field during the Friday night game against La Jolla Country Day. This is where, in just a few months, he would be handing over the school colors to the new incoming president. From the field, he could see the new unnamed stadium packed, with all 787 seats filled, leaving standing room only. The stadium had recently been completed as part of the new building project he had launched during his term. The general could retire knowing that he had indeed made a difference to the 102-year-old Academy.

The new unnamed stadium was funded by the generous donation of the Crean Foundation. A banner was draped over the front of the stadium. For the last week, the community had made a considerable effort to keep the name hidden from the eyes of the general. Finally, as he got to the middle of the field, the ruse was revealed and the cloth dropped, revealing the name of the new stadium: Brigadier General Stephen M. Bliss Stadium. The general was shocked, not expecting such a surprise. Almost in tears, he thanked the community for such an honor. The crowd burst into applause as the new stadium was dedicated to the man who had dedicated twelve years of his life to the Academy.

DUFFIELD SPORTS CENTER

The Academy decided that the construction of a new athletic complex would be the first focus of the Master Plan. The proposed multipurpose athletic complex would be built on the east side of Carlsbad Boulevard over the location of the old field house, an old small cottage that once served as a rest stop for cars coming down the old 101. The new complex was to include a new gymnasium, wrestling room, weight room, locker rooms for home and visiting teams and a trophy room. Up until this point, ANA's athletic facilities have been said to be considered among the poorest in the county, with accommodations so bad that the campus could not host as many home games as other schools in the league. The basketball team had to use the old court, which was not regulation length, inside Davis Hall, which had no heating system. During the same season, the wrestling team had to use the drama stage to practice when the basketball court was in use. Eventually, it was forced to use the outdoor tennis courts for practice, which would then interfere with the tennis team practice. The old weight room was housed where the indoor rifle range used to be—a damp, enclosed space in the basement of Davis Hall with very poor ventilation. This new building would solve all these issues.

Maffucci Field doubles as the multipurpose sports field and parade grounds, thus it is often filled with numerous holes when cadets pivot during parade marches, making for uneven surfaces not ideal for official competitions. The field has no lighting systems, so it is unusable after sunset—this is the reason why ANA could never host a night or CIF playoff game. A lighting installation was part of this Master Plan. The new low-maintenance synthetic turf was to replace the old grass field, cutting future upkeep costs. Surrounding the field would be a new stadium, which would house a press box, concession stand, restrooms, announcer box, VIP box and private boxes on the upper levels for both the home and visiting coaches. This new stadium would also offer better conditions for those in attendance for parades, perfect for the Annual Veterans Day Parade.

As a nonprofit school, money for the building project needed to be raised through alumni and donations. The Academy was just lucky enough to receive a generous donation from the Duffield family, one of the largest in her history. The Duffield family, who had two cadets enrolled—Mathew (Class of 2013) and Kevin (Class of 2018)—donated the initial funds needed for the project.

The groundbreaking ceremony for the new 27,250-square-foot Duffield Sports Center took place in October 2012, with countless alumni, cadets, the Board and city officials in attendance. Brand-new gold ceremonial shovels were lined up for the occasion alongside brilliant white hardhats bearing the logo of ANA. Chairman of the Board Jack Wyatt, cadet Kevin Duffield and Carlsbad mayor pro tem Ann Kulchin were joined by members of the community and key stakeholders as they all simultaneously struck their gold shovels into the soil, launching the start of construction for the large Master Plan.

END OF AN ERA

In 2012, General Bliss was named that year's president of the ninety-eight-year-old Association of Military Colleges and Schools of the United States. That same year, Bliss announced his intention to retire, and the search for a new president was underway.

As the outgoing president, Bliss made sure to finish his twelve-year term on a strong note, seeing out the final stages of ANA's modernization and the completion of the new Duffield Sports Center. As part of the modernization process, around noon on May 31, 2013, a brief ceremony was held in the Information Technology Department as Colonel David Witwer and Director of Finance Janet Pullen pulled the plug on the last remaining MAS90 that once powered the Academy's financial and technology system, the last remnants of the old era.

By June 2013, the Board of Trustees finally selected a replacement for Bliss after a year-long search. The new man would take office early in 2014.

In August, the Army and Navy Academy Television (ANATV) program was launched with a brand-new studio and editing bay located beneath Davis Hall. The new program offered cadets lessons in camera work, news reporting and post-production work. The cadets of ANATV report on news from around campus and the community. By cadet hands alone, they edit and broadcast news stories to the entire community. One of their major stories included coverage of the opening of the Duffield Sports Center.

On October 4, 2013, during the Friday night home game against La Jolla Country Day, General Bliss was called on the field during halftime, where the community surprised the outgoing president by revealing the new stadium was named in his honor.

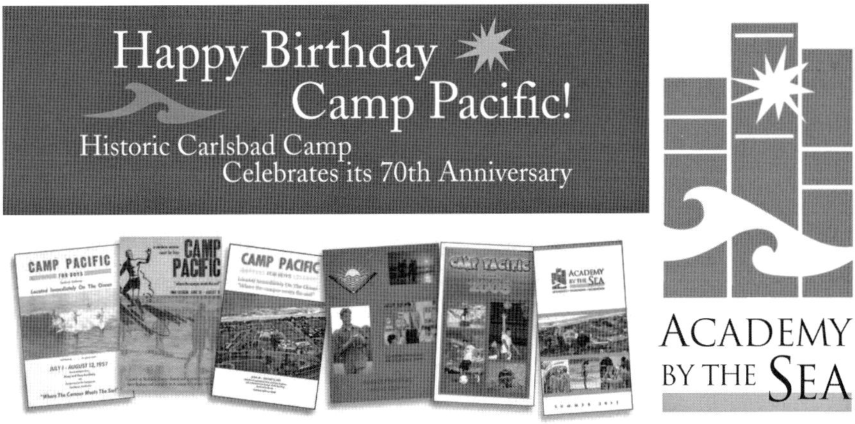

Advert celebrating Camp Pacific's seventieth anniversary. The summer program is also known as the Academy by the Sea. *Courtesy of ANA's Office of Advancement.*

The following day, the new Duffield Sports Center was christened to a cheering crowd of one thousand in attendance. General Bliss led the countdown as the chairman of the Board, Jack Wyatt, and Amy Zeifang, member of the Duffield family and director of Duffield Family Foundation's philanthropic arm, wielded ceremonial scissors to cut the ribbon of the new sports complex. Later that day saw the dedication of the new basketball court.

The Army and Navy Academy Warrior basketball team just finished a strong season, being named the 2013 CIF San Diego Section Champions. To honor the event, the new basketball court was named Court of Champions. Originally, the CIF (the governing body of the California Scholastics Athletics) placed classifying schools in different divisions based on enrollment size. During this period, it did away with the old system and placed the Army and Navy Academy Warrior Basketball team in Division I based on the team's impressive record over the previous five seasons. Cadet Brad Salka, a member of the 2013 team and son of Susan and Scott Salka, who were donors of the new court, took the commemorative first shot into the new basketball hoop. General Bliss accepted the new CIF Championship Banner, which was hung in the new court, and then was presented with a basketball signed by all the members of the 2013 championship team as a token of appreciation to the outgoing president.

ARTHUR M. BARTELL

Major General Arthur M. Bartell was announced in a press release as the new president of the 105-year-old Army and Navy Academy. The major general is a highly decorated U.S. Army officer whose medals for his service in Southwest Asia, Afghanistan, Iraq, Saudi Arabia and the Global War on Terrorism include a Distinguished Service Medal with Oak Leaf Cluster and Defense Superior Service Medal with two Oak Leaf Clusters.

Major General Bartell graduated in 1977 from the University of Michigan, where he was commissioned a field artillery officer after completing the university's ROTC program. He earned one master's degree in behavioral science from Catholic University and another one in national security and strategic studies from the U.S. Naval War College. Bartell had just retired from the army the September before after a thirty-six-year-long career, which included serving as the Commanding General of the U.S. Army Cadet Command, the body that oversees the ROTC and JROTC programs. He was later made a member of the AMCSUS Executive Committee.

"In addition to his diverse military background, Major General Bartell's experience leading the Army's Cadet Command that serves university, college and secondary school ROTC programs across the nation will be immensely important to the quality of the education our cadets receive," added Board Chairman Jack Wyatt in a press release.

On January 17, 2014, the Army and Navy Academy formally bid Brigadier General Bliss a fond farewell in a handover ceremony. The ceremony mirrored the exchanging of Academy colors by the graduating and new Battalion Commanders at the conclusion of the academic year. For the outgoing president, this was a fitting way to welcome the incoming president. This was the first time in the Academy's history that such ceremony could take place, as previous presidencies had ended with abrupt departures or sudden deaths.

Cadets paraded on the field for this first presidential Change of Command Ceremony. The outgoing president passed ANA colors over to the chairman of the Board, who handed the colors to the incoming president. The two generals shook hands, marking a start of a new era.

Major General Bartell addressed the community. "I am extremely honored to be the Academy's new President. I will work tirelessly to build on the accomplishments of General Bliss…to provide the very best academic, athletic, and leadership opportunities." He then concluded his first address to the Corps with ANA's battle cry, "Warrior Pride!" then added a new cry, "Warrior Strong!"

In an outgoing interview, Brigadier General Bliss stated, "When I became President, I told the Cadets and faculty at our first character development session that anything worth doing is worth doing well," noting that he hoped he had left a strong legacy of professionalism in education and an environment of respect among the cadets. Bliss concluded, "One thing is for sure, we will never forget our time here and the many wonderful people who have made the Army Navy experience such a great one for us."

Old Soldiers Never Die

The Golden Age of the American Military School has long since passed, and it is now in its final twilight hours. The few survivors of this shrinking community soldier on, as more of their peers, without observance, quietly close their doors. As noted earlier, the rise of the JROTC program has greatly affected the creation and continued existence of remaining military schools. Originally, the JROTC program was meant as an alternative for students who could not afford to enroll in a military school; now it has become the default system for all pre-college military education. Yet, as the War on Terror trudges on, paired with downsizing in the military making the need for new soldiers and officers less of a priority, the JROTC and ROTC programs have also seen a dwindling popularity. With opposition on all fronts, including groups campaigning for its defunding, like the Coalition for Alternatives to Militarism in Our Schools and Project on Youth and Non-Military Opportunities, the JROTC program is currently facing negative backlash in the public conscience for "militarizing" children just as the military school community experienced decades previously.

The state of remaining military schools is even more disheartening. By this point, the AMCSUS membership had dwindled to just twenty-eight pre-college military schools. Even major players in the military school community have since decreased in prominence. In the neighborhood of West Point, the New York Military Academy (1889) was forced to close in 2010 due to low funding, despite having an alumni base that includes business mogul and president Donald Trump. It later reopened due to alumni support with a dismal student body of eighty before filing for bankruptcy and closing in 2015. Marmion Military Academy (1933) dropped its military portion, while La Salle Military Academy (1883) and Kemper Military School (1844) filed for bankruptcy and closed.

Other military schools did not fare better. If not closing, most see an enrollment of fewer than one hundred cadets. The oldest surviving military school, Carson Long Military Academy (1836), has an enrollment of ninety cadets. In a 2014 radio report, Oak Ridge Military Academy (1852) was said to have gone down from two hundred to forty-five cadets. Howe Military Academy (1884) was last reported to have a corps of seventy-four cadets. And Chamberlain-Hunt Academy (1879), one of the oldest college preparatory schools in the South, founded as Oakland College in 1830, was last reported to have a meager corps of four cadets, not even enough for a single squad, before announcing its closure in 2014.[37]

As of 2014, the AMCSUS members with the highest enrollments are joint high school/college establishments, like Valley Forge Military Academy and College (1928) with 602 cadets and Wentworth Military Academy and College (1880) with 1,500 cadets, as well as state-sponsored schools like New Mexico Military Institute (1891), with 914 cadets, and Georgia Military College (1879), with 8,000 cadets. However, these numbers are somewhat inflated, as they combine the student body from all sectors, with students from the college or junior college portion making up the largest percentage. Wentworth Military Academy's enrollment is mostly made up of commuter civilian students not in the military program. Its combined high school and junior college Cadet Corps numbers at only 250. Georgia Military College's large number shrinks when revealed that its high school enrollment is also at 250 cadets.[38]

For just pre-college military schools, the Army and Navy Academy, with an enrollment of around 350 cadets, sits nicely with other steady military high schools like Marion Military Academy (409 cadets), St. John's Northwestern (300 cadets), Camden Military Academy (302 cadets), Randolph-Macon Academy (350 cadets), Fork Union Military Academy (463 cadets) and Riverside Military Academy (475 cadets). With enrollment numbers for these schools stabilizing after the recession, it is safe to say that

37. These numbers are based on the most recent account available during the time of research, 2014. The actual numbers may be different than the ones publically reported and may have changed since publication of this book. Enrollment reports are cited in the Additional Source Notes.

38. Although schools like Wentworth Military Academy and Georgia Military College have the largest reported number of cadets, when taking into account the high school cadet size, they are just under the enrollment size of strictly pre-college military schools, which range from three hundred to five hundred cadets.

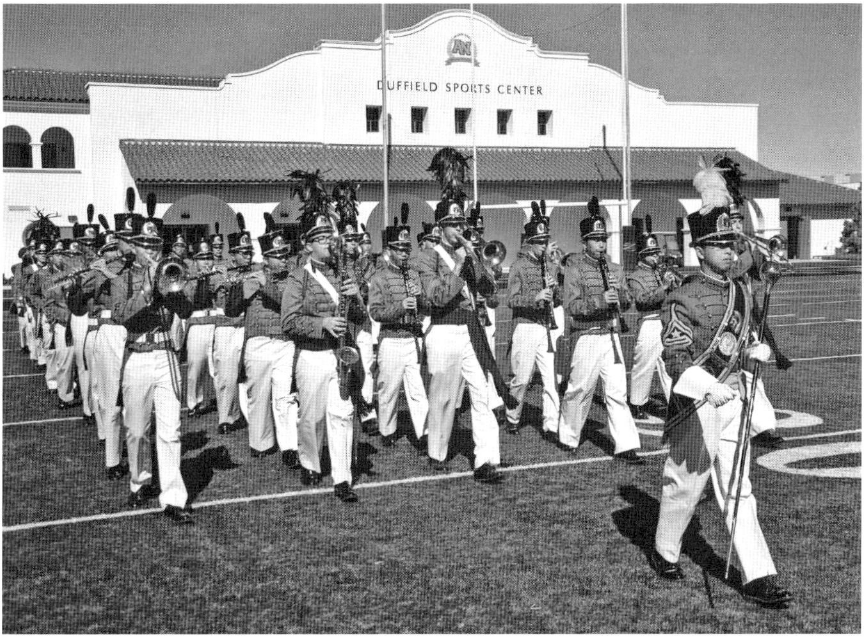

The Atkinson Warrior Band leading the parade at the new Duffield Sports Center Complex. *Courtesy of ANA's Office of Advancement.*

the Army and Navy Academy and her peer pre-college military schools are not going away anytime soon—never at the same prominence once held in national culture however. The United States is the only nation that has ever given rise to a community of private military schools at such a level. These endangered institutions are as American as baseball and apple pie, but they have long since passed their prime, only to be remembered in culture as relics of a bygone era.

In light of all this, the Army and Navy Academy stands strong as a beacon of what such a unique institution can offer to the community and the nation. While other military schools appear to be on their last legs, ANA is excelling in her second century of existence.

Back to Basics

Within months of Bartell becoming the new president, it appears that most of the post-2008 recession changes have been reversed. ANA has reverted to its traditional High School and Middle School structure, RLOs have returned to being TAC Officers and longtime faculty are being promoted, while the Centennial logo is slowly being phased out.

Focus has been shifted from trying to emulate other schools towards working with the individual traits that make ANA distinct from her peers. This is what Davis had always intended; he knew that what worked for one school would not work for another, and trying to shoehorn such a structure would dilute any functionality. Davis would always experiment with his own teaching structure. The Davis System was always evolving, utilizing new ways to better educate students and serve the community.

With this same mission in mind, in 2014, the Army and Navy Academy partnered with the Gurian Institute to incorporate new teaching techniques on the forefront of the educational field. As opposed to hour-long lectures, ANA shifted towards progressive teaching methods and customized lesson plans scientifically aimed to the way young male brains are wired. This included making use of "brain breaks," graphics and visuals in teaching, movement around the classroom and more hands-on learning. The school also strongly emphasized the value of small teacher-student ratios and the benefits of boarding environments for college preparation moving forwards.[39]

In recent years, the humble little Astronomy Club that I tried to revive as a cadet has grown to greater prominence. The Army and Navy Academy is now the only high school in the state to offer a University of California–accredited astronomy course, which now works with the Harvard College Observatory and Johns Hopkins Applied Physics Lab and is named a campus branch of the Institute for Student Astronomical Research. The cadets of the Astronomy Club have since discovered three new asteroids and had a part in writing four peer-reviewed research papers submitted for publication.

Along with the Annual Veterans Day Parade and the Atkinson Warrior Band's return to high distinction, ANA even rose to greater prominence within the community for taking part in the Seventieth Anniversary of D-Day Commemoration. On June 6, 2014, a select group of thirty-one cadets arrived in Normandy, France, to participate in the commemoration events. The Army and Navy Academy was the only school in California invited to send a contingent to march in the D-Day Anniversary Parade and Music Salute. The City of Carlsbad recognized the Army and Navy

39. The Academy was named by Michael Gurian, founder of the Gurian Institute, as the Tenth Gurian Institute Model School in the nation.

Academy for being selected as California's sole representative during the commemoration, with Mayor Matt Hall memorializing the event by proclaiming April 23, 2014, as Corps of Cadets Day.

During the first one hundred days of his presidency, Major General Bartell utilized his experienced background and introduced many new improvements ensuring that the Academy would continue on this high for the next century. While it is still too early to determine the impact of his presidency, it is safe to say that with her strong alumni base and community of patrons, teachers and cadets, the future of the Army and Navy Academy is in safe hands.

In the 2014 issue of *Army Navy*, the new president concluded his address to the Corps with, "The first 100 days have been exciting and fulfilling, and the future promises to be even more so. We are enjoying every minute of this new adventure. Warrior Pride! Warrior Strong!"

AFTERWORD

STORY OF A CADET LIFER

My life on campus started in 1967 at 11 years of age when I was enrolled to Camp Pacific and ended with graduation in June of 1972 at 17 years old. I was not the only one to attend the Camp and find out towards the end of a carefree summer program that my parents had enrolled me to attend school there in just a few weeks. Camp was great and Coach Maffucci was fantastic with all of the kids. He cared. I was chubby and young, but competitive enough to be awarded the Best Sportsmanship award for the summer. To help the transition, while delivering me to Military School, my parents stopped in Dana Point at Hobie Alder's original surf shop. We talked to Hobie and Corky Carroll and ended up buying a 6'8" Corky Carroll super mini board. What a treat and piece of history.

Junior School was a shock. Nothing like Camp Pacific. However, under the guidance of Major David Stoll, we all survived. He was my main mentor and father figure during those 2 years. Feeling abandoned by my parents, this man really cared for me and it was a blessing he was there. I learned marching, drilling, and general discipline early on, which gave me and the other cadets who continued to upper division, a head start.

The other strong influence in my life was Tennis Coach Harold Moses. I played for the team my entire time at the Academy. He was another one who really cared about what was happening on the court and what was going on in your head. He would take you off campus to buy good "Converse" tennis shoes, and also offered fatherly advice through my high school years. He essentially took over where Stoll left off. They appeared to be friends and

I'd sometimes see them talking while looking my way. They had my back! I learned to love the game of tennis and managed to stay #1 most of the time, which made Coach Moses happy. He suffered some terrible disabilities from war time injuries which many other cadets were never aware of, but he could still beat me when he wanted to prove a point. A true Warrior indeed.

I participated in other sports, but nothing interested me like tennis. I guess I thrive in one on one competition. You would think with my history with Coach Maffucci, that I would have played football throughout the years. I certainly had the physical build for the game. However I got soured on the prospect from the JV coach with his opening speech. He essentially said that we had to be willing to break bones, get knocked out, and endure painful training sessions. He wanted us to give our life for the sport. Further, if anyone didn't want to do that, turn in your gear right now, which is exactly what I did! He should have saved that speech for the college teams, which is probably where he heard it. Coach Maffucci was always a bit disappointed about that, but he knew I was supporting the school sports program and winning tennis matches. Winning was not always the case unfortunately, and certainly not a common occurrence with our small school.

Colonel Atkinson was a Grandfather to us all. We listened to his Chapel sermons about our lives on a very regular basis. He believed in what he was saying and we in turn believed him. I got to know him on a more social basis after graduation that just wasn't possible as a cadet. I didn't know of anyone who was able to be as close to him as we had been with our other mentors. I did have a few visits to him in his upstairs office in Davis Hall. One meeting was early on for something I did wrong and the other formal meeting was many years later congratulating me for my successes and being #1 in my graduation class. I was on top of the world. This fine gentleman was giving me a compliment! Many cadets avoided him when they saw him walking on campus, but I never did. It was comforting that he also knew who we were over the years and we felt like family.

Dean Ede was yet another one of those special people who remembered who you were and had your GPA in his head. Most of us spoke to him more often than we did the President, Col. Atty. He helped with our college searches and provided advice about the courses necessary to fulfill requirements of colleges. These classes were offered at the academy summer school. It seemed to me that he had a good sense of humor, but he didn't often show it. He was nice enough to share information that he wasn't always authorized to do. Many of us were very curious about our IQ test results. I visited him to ask him and he said he couldn't tell me.

Then he pulled open a file cabinet drawer and dropped a paper in front of me and said he would be right back. I read the results and then returned the paper facedown. When he returned, he asked if there was anything else he could help me with. I thanked him and said no.

My high school years were productive and yielded mixed emotions. I knew the system and thrived. I was promoted regularly, won lots of merit badges and awards, and ultimately became an officer. I was especially proud of being a part of Warrior Newsletter production which John Gianelle (Parsons) and I edited. However, we were not necessarily happy during the changing times.

Compared to my contemporaries, walking around campus with a cap and a sword at your side was otherworldly. The other side of the coin was being away from home and not being able to do the things our friends there were doing. The antiwar movement was popular, and every one of my contemporaries in my hometown of Claremont, California, had long hair and wild clothes. It was not the best time socially to be a ROTC cadet. One thing that no one else I knew had, and they couldn't ever achieve, was living at and with the beach.

I'd like to mention Joe from the Mess Hall. He was there when I started. He too, was a young guy, no doubt a few years older than myself to be working full time. He is still working as of this writing, and is a really nice humble guy. The cadets of my era had a bad habit of throwing their finished food trays with plates across the counter upon leaving. Many times with too much velocity while yelling "Hey Joe" as the plates were hitting him.

I've got to thank and credit this grand institution for the success I have enjoyed in my lifetime. I surmise that it is 90 percent responsible for the favorable outcome of my life. The love, the discipline, the patience, and the caring that was provided to me, was incomparable. I applaud those who are willing to fill those shoes and "connect" to the young guy in front of them. I am certain that I would not have achieved the successes in my life without this institution. My goal is to give back many times more than what I received. The ultimate goal for my academy should be to maintain a staff of the type of dedicated individuals that helped me thrive. That's a tall order.

—Bradley D. Larsen, Class of 1972

Appendix I

Presidents of the Academy

Colonel Thomas Alderson Davis (Founder)
November 23, 1910, to 1938
Major John Lynch Davis Jr.
1939 to 1941
Captain V. R. Vestal (Acting President)
1941 to 1942
Major Samuel Warfield Peterson
1942 to April 26, 1943 (died in office)
Colonel William C. Atkinson Sr.
1943 to 1972 (first term)
Lieutenant Colonel William C. Atkinson Jr.
1972 to 1974
Colonel William C. Atkinson Sr.
1974 to 1980 (second term)
Colonel Gilbert N. Brookhart
June 1980 to October 1982 (died in office)
Mr. Fred Rowbotham (Acting President)
October 1982 to August 1983
Colonel Donald H. Cline
August 1, 1983, to January 25, 1991
Dr. Conrad R. Garner
January 25, 1991, to April 11, 1991

Coach John Maffucci (Acting President)
April 11, 1991, to August 21, 1991
Mr. Michael Wallgren
August 21, 1991, to June 9, 1992
Dr. Fred W. Heinle
July 13, 1992, to June 21, 1995

[From June 21, 1995, to October 24, 1998, there was a period with no official president. Three individuals made up a group known as the Office of the President]
Dr. Fred W. Heinle (Headmaster)
June 21, 1995, to June 26, 1997
Johnnie Crean (CEO, Chairman of the Board)
June 21, 1995, to October 24, 1998
Lieutenant Colonel Stephen A. Miller (Superintendent)
June 21, 1995, to October 24, 1998

Lieutenant Colonel Stephen A. Miller
October 24, 1998, to February 8, 2002
Brigadier General Stephen M. Bliss
February 8, 2002, to January 17, 2014
Major General Arthur M. Bartell
January 17, 2014, to present

Appendix II
Songs of the Academy

Alma Mater of the San Diego Army and Navy Academy

This is the alma mater of the San Diego Army and Navy Academy. It was written in 1930 by then bandmaster Colonel Atkinson. It was used from 1930 until 1936, when the Academy moved to Carlsbad.

By the City's Northern Border reared against the Sky,
Proudly stands our Alma Mater as the years roll by.
Forward ever be our watchword Conquer and Prevail,
Hail to thee our Alma Mater A. N. A. all Hail!

Alma Mater of the Army and Navy Academy

The second alma mater of the Army and Navy Academy was written by President Colonel Atkinson and Bandmaster Al Polhamus. The alma mater is still in use today.

Sheltered by the Sun kissed hills, Bordered by the Sea,
Hail to Army Navy, Our mother ever be!
May thy standards lead us on—on to victory.
Hail to Alma Mater, All Hail! All Hail! To Thee.

Proud we stand as one the Corps, proudly do we sing.
Brother stands by brother, now hear our voices ring.
May your standards keep us strong, strong in loyalty,
Hail to Army-Navy, All Hail! All Hail! To Thee!

Loyal Sons of ANA let your voices ring,
Praising Alma Mater, to Her we'll honor bring.
May thy memory linger on through each fleeting year,
Binding us together in pathways far and near.

"HAIL TO ARMY NAVY"

"Hail to Army Navy" is a fight song written by President Colonel Atkinson and Bandmaster Al Polhamus. It was used as the de facto anthem of the Centennial.

Hail to Army Navy! Gather 'round and praise her name.
Hail to Army Navy! Today we're going to win the game,
Rah! Rah! Rah!
Hail to Army Navy! For the black and gold we'll fight.
We will always fight, fight with all our might! Yes, Hail to the Academy!

"MARCH ARMY NAVY MARCH"

"March Army Navy March" is a march written by President Colonel Atkinson and Bandmaster Al Polhamus for the Academy. It has not been used in the modern era.

March Army Navy March, on down the field.
To win a Victory for A.N.A. this game we'll win today.
March Army Navy March, go cross that goal.
For Alma Mater and Auld Lang Syne, drive through that line.

APPENDIX III

FLAGS OF THE ACADEMY

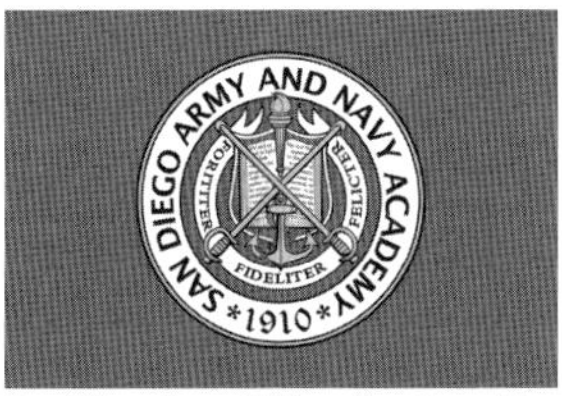

San Diego Army & Navy Academy Flag
Used from 1910 to1936

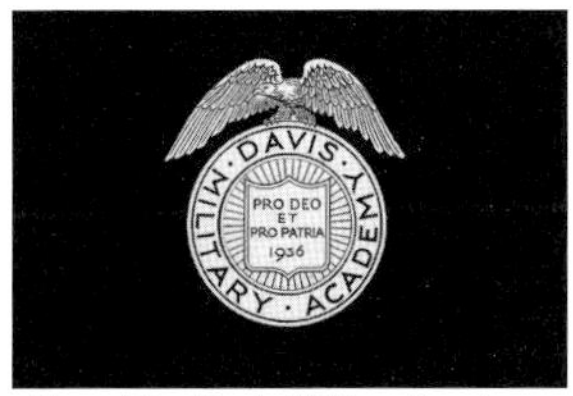

Flag of the Davis Military Academy
Used from 1936 to 1937

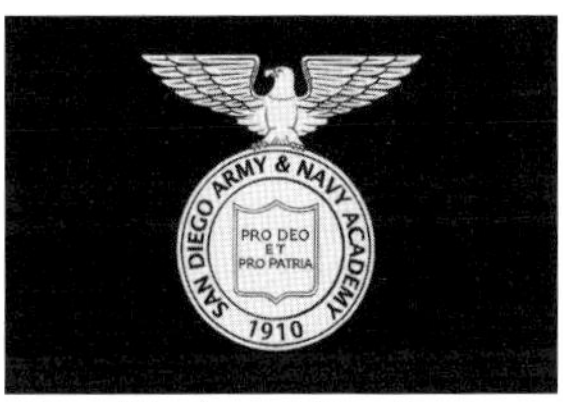

Flag of the Army & Navy Academy
Used from 1937 to 1943

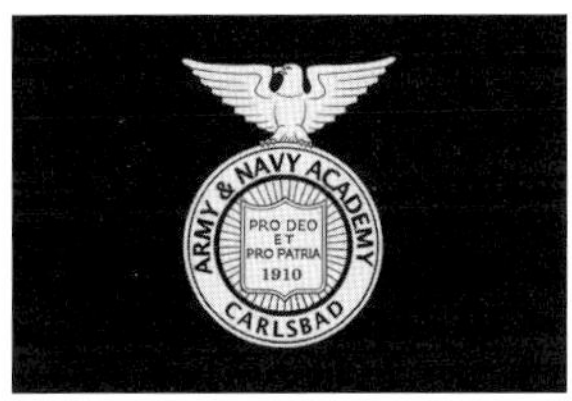

Flag of the Army & Navy Academy
Used from WWII to Vietnam War

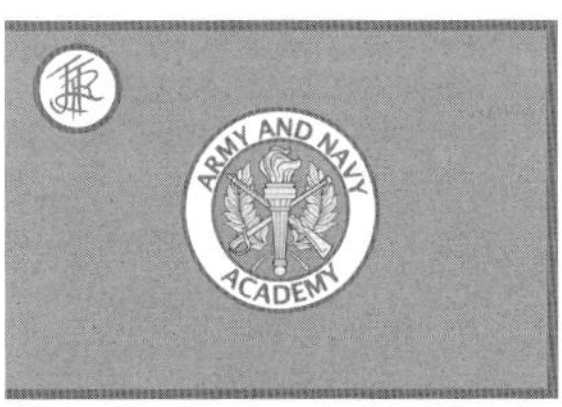

Flag of the Academy's JROTC Unit
Used alongside Academy flag

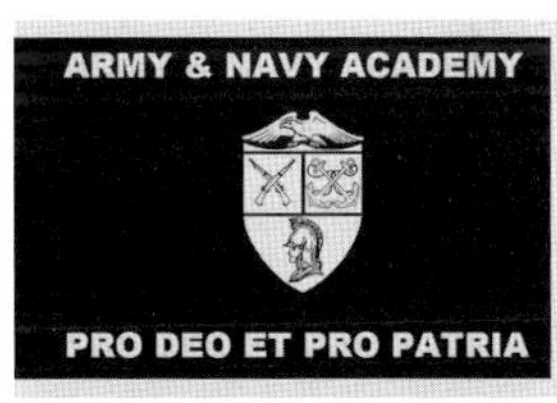

Flag of the Army & Navy Academy
Used from 1980's to 2009

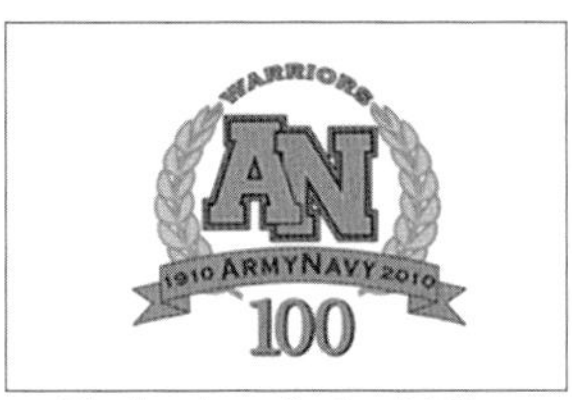

The Academy Centennial Flag
Used from 2009 to 2011

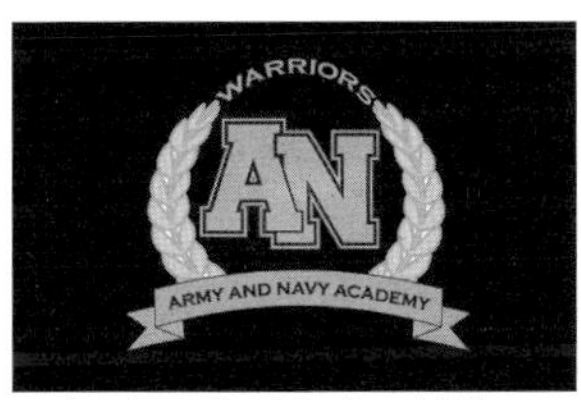

Academy Post-Centennial Flag
Used from 2011 to Present

Flags of the Army and Navy Academy. *Designs re-created digitally by author.*

APPENDIX IV

SEALS AND CRESTS OF THE ACADEMY

Seals of the San Diego Army & Navy Academy (Pacific Beach)

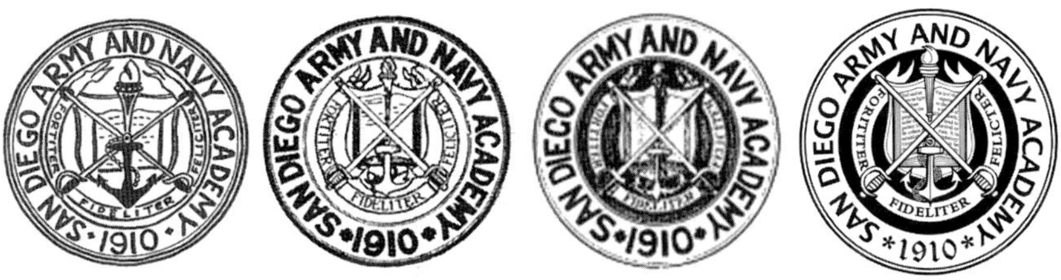

Seals of Davis Military Academy/Army & Navy Academy (Carlsbad)

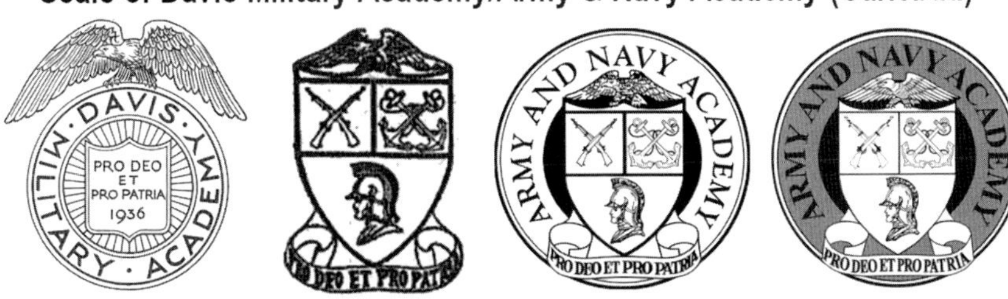

Seals of other Army & Navy Academy Schools and Divisions

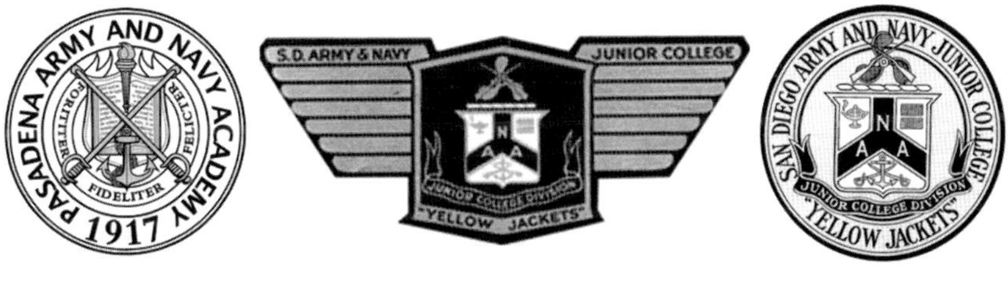

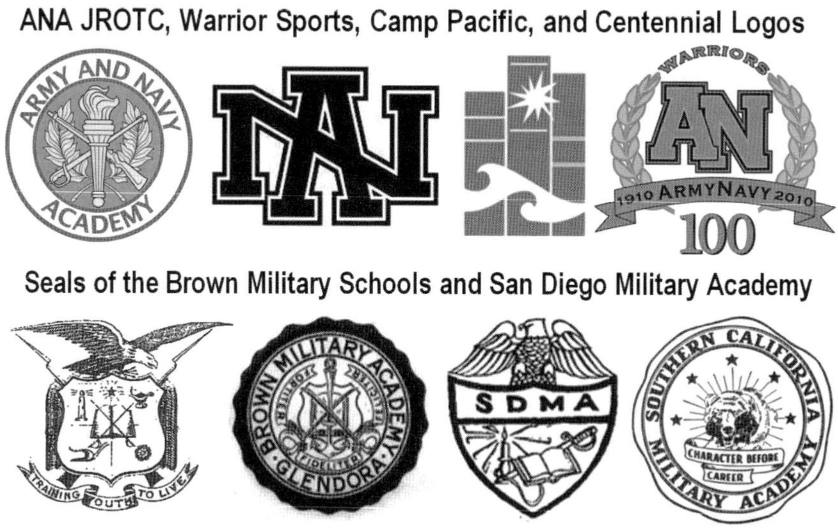

Seals and symbols of the Academy. *Designs re-created digitally by author.*

ADDITIONAL SOURCE NOTES

The following is a referential supplement to the book. As most of the information for the early chapters of this book comes from old newspapers, books out of publication and sources unavailable to the public, I have collected some relevant quotes from these sources to help back up findings made in my research. Information that is readily available to the public, books still in print or articles accessible online are cited but not reproduced.

Most of my information for the early years (1910–13) of the San Diego Army and Navy Academy comes from newspaper adverts, 1911 letters housed in the William C. Atkinson Museum and the 1913 yearbook.

A backbone of the history of the San Diego Army and Navy Academy (1910–36) comes from a document housed in the Carlsbad Library entitled "The Army & Navy Academy: Its Story 1910 through 1936." It is believed to have been written by Colonel William C. Atkinson, as stated in its introduction: "The information presented in this first chapter of the history of the Army & Navy Academy…is the result of several years of research and study over a period of years starting in 1972 by Col. W.C. Atkinson, currently president of the academy. He carefully studied every available old record of every kind; wrote letters for information to numberless former instructors and cadets, and quizzed others who might be of help. He did a tremendous job and everyone interested in the academy, past or present, should appreciate it."

Information on individual presidents came from a document I found in archives of the Army and Navy Academy in 2006 entitled "History on Select Presidents of the Army & Navy Academy," specifically concerning the presidencies of John Lynch Davis Jr., Captain V.R. Vestal, Major Samuel Warfield Peterson (1939–43) and Colonel Donald H. Cline (1983–91). This document helps provide the backbone of chapters 9 and 15.

Information concerning the Brown Military Academy and San Diego Military Academy come from the book *The Four Academies* by Louis T. Bitterlin (Superintendent of Brown Military Academy and founder of San Diego Military Academy) and Milan Radovich, published in 1996.

Some of the information from chapters 16 and 17 comes from interviews I conduced with Johnnie Crean (Board member, 1991–2003), Karin Chemlir Miller (wife of Lieutenant Colonel Stephen Miller) and General Stephen Bliss (president, 2003–14). As these interviews are not available publically, I have included large selections of these interviews in their entirety so the context of the information and quotes are available for assessment. Due to space concerns, I had to cut down the size of the Additional Source Notes in this printed version; a more detailed version will be available at armynavyacademy.tumblr.com.

Prologue: Marching Westward

Introduction: The Military Schools of America

Military Uniforms in America. Vol. 2, *Years of Growth, 1796–1851.* "Cadets, U.S. Military Academy, 1816–1817." N.p.: Company of Military Historians, 1977.

Forman, Sidney. *West Point: A History of the United States Military Academy*. New York: Columbia University, 1950.

Depuy, R. Ernest. "Mutiny at West Point." *American Heritage*. N.p.: American Heritage Publishing, 1955.

Ellis, William Arba. *Norwich University: Her History, Her Graduates, Her Roll of Honor.* N.p., 1898.

Holden, Russell J. *Norwich University Cadet Handbook*. Office of the Commandant. Northfield, VT: Norwich University, 2013.

Andrew, Rod, Jr. *Long Gray Lines: The Southern Military School Tradition, 1839–1915.* Chapel Hill: University of North Carolina Press, 2004.

Chapter 1: Sons of the South

Information on Thomas A. Davis has been pieced together from various sources, including the following: *University of Tennessee Record.* Vol. 1. Knoxville: University of Tennessee, 1897; *San Diego County California.* Vol. 2, *A Record of Settlement, Organization, Progress and Achievement.* N.p.: S.J. Clarke, 1913; McGroarty, John S. *Los Angeles from the Mountains to the Sea.* Vol. 3, *Select Biography of Actors and Witnesses to the Period of Growth and Achievement.* Los Angeles, CA: American Historical Society, 1921.

From a newspaper clipping entitled "Facts Pertaining to the San Diego Army & Navy Academy and Its Work," circa 1911: "The record of the Superintendent, Capt. Thos. A. Davis, is as follows: B.A., University of Tennessee, 1890–1891, 1891–1892; First Lieutenant Sixth United States Vol. Infantry, June 29, 1898; Regimental Adjutant, Sixth United States Vol. Infantry, Dec. 1, 1989; Captain, Sixth United States Vol. Infantry, Jan. 7, 1899. Served in Porto Rico during the Spanish-American War."

From "The Army & Navy Academy: Its Story 1910 through 1936," by Colonel William C. Atkinson, 1972: "The founder of the Army & Navy Academy was a remarkable man; Col Thomas A. Davis was raised in a Christian family in Tennessee, and developed into a young man of the highest character. He was a top student in high school and a pitcher on the baseball team. He graduated from the University of Tennessee in 1890 as a highly regarded student and also found time to continue his pitching; sing in the Glee Club; played a cello in the orchestra, and was captain of the battalion of cadets for military training. During the Spanish-American War he was a captain in the 6th U.S. Volunteer Infantry. Following the war he went into business with his father but a growing desire to teach young men and influence their lives."

Most of the information on the El Paso Military Institute comes from Samuel A. Ramirez, "El Paso Military Institute," *Junior Historian* 14, no. 5 (March 1954), from the Texas State Historical Association: "The school song was called Sons of the South, and was dedicated to the military institute. The original version of the words was first sung at the Sewanee-Vanderbilt game in Nashville on Thanksgiving Day, 1908. The music and words were published by the Pass City Publishing Company of El Paso in 1911."…"Cadets also took part in parades in the city. In 1909, when President William H. Taft and General Porfirio Díaz met in El Paso, the cadets of EPMI were one of the main attractions in the parade."… "During World War I some of the ex-students of EPMI were taken into the service. James B. 'Buster' Biggs, one of the former cadets, served as a lieutenant in the United States Air Force and was killed in a flying accident in Europe in 1919. Biggs Air Force Base in El Paso was named for him."

"El Paso Military Institute Site at Fort Bliss," article published on the UTEP Library Special Collections website, was also used.

Borunda, Daniel. "First Meeting of US, Mexico Leaders: Museums, Events Mark Centennial." *El Paso Times*, October 19, 2009.

PART I: BY THE CITY'S NORTHERN BORDER

Chapter 2: Founding of the Academy

From the foreword of *The Cadet*, 1913, Thomas A. Davis: "Attracted by the location of San Diego, its unequaled climatic conditions and what I had heard of its past, so rich in history and romance; of its present with its wonderful—almost unprecedented—growth and improvement; and what I believed the future to have in store for the city, I yielded to the request of some of its citizens and determined to establish here a high grade military school. So on November 23rd, 1910, the opening exercises of the San Diego Army and Navy Academy were held in the auditorium of the large buildings which had been known previously as the Hotel Balboa, located in the center of Pacific Beach, the delightful suburb of San Diego.... The names of the following boys were mentioned as the cadets enrolled: W.D. Cannon Jr., William Fox, Harlow Garetson, Willard B. Hage, Sterling Hebbard, Forrest Henry Hieatt, Rex Minton Hocker, Charles Kingsley Johnson, James Hurvey Johnson, James Parker, Stewart McKee, Guy Parmenter, Wm. Armstrong Shaw—the 'immortal thirteen,' in this case, a lucky number, as I had only expected ten, because of the lateness of the date of opening."

Scenic America. "A Select School For Manly Boys." Circa 1910s.

Newspaper clipping entitled "Facts Pertaining to the San Diego Army & Navy Academy and Its Work," circa 1911.

In the fiftieth anniversary edition of *The Warrior*, it notes the first football team with thirteen cadets in 1911 and quotes *The Cadet*: "Although a young school and older schools can beat us in football, suppose we have competitive drill and see who holds the long end of the score?"

Chapter 3: Those Distant Years

From *The Cadet*, 1913, Thomas A. Davis: "When the examiner from the University of California was shown the plans used by the Academy, whereby the gradings and general record of each boy are kept under the observation of every member of the faculty, and the methods for securing effective work were explained, he said, 'In my ten years' experience, I have not seen a better system for looking after the individual student.'"

From "The Army & Navy Academy: Its Story 1910 through 1936," by Colonel William C. Atkinson, 1972: "The third year of the life of the Academy saw a tremendous growth in the number of cadets, the faculty, and the activities. With 150 cadets on the campus it was possible to organize four cadet companies and a band which also became an orchestra. This led the Colonel to strengthen his belief that he would develop the largest military preparative school west of the Mississippi.…One of the highlight of that year was the official recognition and accreditation of the school by the University of California.…Examiner of Schools in the United States gave the Academy the Highest Scholastic rating under examination in the United States, and the Federal Inspection Recorder gave it a very high record.…It is mighty interesting to learn that 72 cadets had served their country in the war effort and that 20 were commissioned officers."

Clarke, S.J. *San Diego County California—A Record of Settlement, Organization, Progress and Achievement.* Vol. 2. Chicago: S.J. Clarke, 1913.

Chapter 4: The Lost Academy

A majority of the information in this chapter comes from various newspaper advertisements concerning the Pasadena Army and Navy Academy and Pasadena Military Academy.

From McGroarty, John Steven. *Los Angeles from the Mountains to the Sea.* Vol. 3. New York: American Historical Society, 1921: "Pasadena Military Academy owes its birth and healthy growth solely to the determined efforts and executive ability of one woman, Mrs. John H. Henry.…Having a boy of her own for whom she had definite ideas of schooling and not wanting to send him from home, she finally determined to put her ideals into concrete form and build a school such as she had in mind for her own son.…Her ideal school was a military character, but it was to be more than a formal military school, and equally indispensable were the sweet home influences and the moral atmosphere which are so notably absent in many military academies. Other things upon which she placed great emphasis in her plans was a perfect cuisine, a clean dormitory, and a wholesome environment in every respect.…The school was opened October 8, 1917, with an original enrollment of eighteen pupils."…"To leave nothing undone that may complete her ideal, Mrs. Henry contemplates extensive improvements in equipment and buildings, involving expenditures of about sixty thousand dollars.…Mrs. Henry's ambition is to make the Pasadena Military Academy the finest school of its kind in the West, and demonstrated results of her first experience and her well-known determination make that achievement hardly a matter of doubt."

From Sargent, Porter. *A Handbook of American Private Schools*. 7th ed. Sargent's Handbook Series. Boston: Cosmos Press, 1922: "[N]ow in its fifth successful year.…The school is sanely military. Care is taken…that cadets do not exercise beyond their years."

PART II: THE ACADEMY'S GOLDEN AGE

Chapter 5: California's West Point

The name "West Point of the West" for most of the twentieth century in press and media has always referred to the Army and Navy Academy, which was the first to be identified with this title in the 1920s. Other schools that have also used this nickname afterward include Brown Military Academy after purchasing the Pacific Beach campus, New Mexico Military Institute, Utah State University ROTC and Oregon State University ROTC. "West Point of the West" was used as the name for the centennial documentary for the Army and Navy Academy in 2009.

From "The Army & Navy Academy: Its Story 1910 through 1936," by Colonel William C. Atkinson, 1972: "As the Academy grew in numbers and recognition it added to Col. Davis' deep seated dream that his school would surely become the largest of its kind west of the Mississippi and rated as the highest military preparative school in the United States. Probably the most important development of the year was its entrance into naval training in a real way to completely round out its field of effort and accomplishment in every phase."

From the fiftieth anniversary edition of *The Warrior*: "[D]ances called Hops in West Point fashion, rally before big game very enthusiastic, one with Loyola ended when a 'long line was formed with the corps in their pajamas, and they serpentine down to the Crystal Pier Ballroom, where some yells were given. The bunch then trailed back to the school, holding up traffic all the way.'"

Los Angeles Times. "Pacific Coast Eleven to Be Named on Saturday Preceding Thanksgiving Day." November 7, 1922.

City of San Diego and San Diego County—The Birthplace of California. Vol. 1. San Diego, CA: American Historical Society, 1922.

Chapter 6: Conquer and Prevail

From "The Army & Navy Academy: Its Story 1910 through 1936," by Colonel William C. Atkinson, 1972: "(1926–1927) In the yearly catalogue that year the Colonel stated that his school was now 'The West Point of the West.' That year the Academy joined the Association of Military Colleges and Schools of the United States."…"(1927–1928) Special recognition for the year came when the entire corps was transported by train to Los Angeles to take part in the parade that dedicated the new Los Angeles City Hall, and furthermore, was awarded the Silver Cup as the best military unit in the parade."…"(1929–1930) The highlight of the year was the invitation for the corps to lead the Rose Parade on New Year's Day in Pasadena which means the opportunity to see the Rose Bowl Game as well."…"(1931–1932) This year saw the cadet corps reduced further to only 185, and the administration of Colonel and Major Davis, sole owners of the property, were forced to reduce all salaries in the effort to meet obligations, and every effort was made to keep expenses to a minimum."

From the fiftieth anniversary edition of *The Warrior*, published in 1960: "[D]uring the Los Angeles City Hall dedication, the Academy won a cup for excellence in drill, discipline, appearance, and military bearing."… "New years day, 'as a result of the committee in charge, the cadets marched in the Rose Tournament Parade.' Catalogue for academy said 'The training value of such a trip is well worth while, as it impresses on cadets the necessity for thorough preparation, organization, and control.'"

From "Let's Get Acquainted," an advert created by Colonel Atkinson promoting the Academy: "On New Year's Day of 1930, I was privileged to witness the most beautiful sight of my years to that date; namely, that of seeing my Cadet Corps of 550 boys in their West Point dress uniforms leading the Rose Tournament Parade.…[N]ever did I dream that one day my boys would be leading that world-famous parade."

From *Montana Standard*. "Gen. Forsythe." August 31, 1969: "Gen. Forsythe was commissioned a second lieutenant in the regular infantry while serving on Kodiak Island in February 1942....He graduated from Butte High School in 1935, San Diego Army and Navy Academy Junior College in 1937."

From Forsythe obituary, *The Item*, December 30, 1987: "Lt. Gen. Forsythe, described by retired Gen. William Westmoreland as 'the father of the modern Army,' has died at the age of 69."

Chapter 7: The Black Year

From Turner Classic Movies article on *Dinky* and Mary Astor: "A scene in which *Dinky* gets into a fight during football practice was shot on location at a military school, with actual students at the school. Cooper insisted that the fisticuffs be real, and was frustrated and angry that the other boys weren't really punching him. Astor writes that he ran to his mother crying, 'I hate being a movie actor—I hate it! I want to be a guy, a man; I don't want fellows like this to treat me like a sissy!'"

From *The Cadet*, 1936, yearbook published by the cadets of the San Diego Army and Navy Academy: "In choosing Miller at halfback, the selection board made no mistake. Chosen almost unanimously, Kess was the one outstanding candidate for a backfield position. Throughout this section of the country, Miller was heralded as the greatest passer seen in some time. In six games he had 73 of his accurately thrown passes completed. His blocking and running ability were without an equal in the Junior College ranks."

From "The Army & Navy Academy: Its Story 1910 through 1936," by Colonel William C. Atkinson, 1972: "In the fall of 1935 the student body had grown to 225 cadets, and that year was one of the finest ever. The Junior College Football Team listed five members on the 'All Southwestern Players'; the band and orchestra had a banner year topped with again serving in the Pasadena Rose Parade....Davis, minority members of the Board of Trustees, received official notice that the bank and other creditors intended to take over the entire property and all of the assets of the Academy. It is impossible to realize what this meant to them and the many devoted people who had spent years in building one of the highest regarded preparatory schools in the country."

From Bitterlin and Radovich, *Four Academies*: "The Security Trust and Savings Bank took over the ownership of the Academy. The Colonel continued to head the school. The public never really knew of the actual circumstances, which now existed."…"They notified not only the public, but their former and present students and their parents, of their intention of starting a new school in Carlsbad.…The press had a field day with this turn of events. The newspapers stated that the Colonel had been asked by The Security and Trust Bank to surrendered much of his personal control of the institution. The Bank's simple press release was, 'Influences of the Depression led to lower enrollment and there were not enough funds to pay expenses associated with the building projects.'"

Part III: Sheltered by the Sun-Kissed Hills

Chapter 8: Bordered by the Sea

From Thomas A. Davis's letter to V.J. Dorman on November 17, 1937, part of "Davis Academy Formal Opening Well Attended," unknown paper, September 24, 1936: "My dear Mr. Dorman, When in the city a few days ago, I tried to reach you by calling at your office buy you were not in, and when I tried your residence, the telephone was busy. I had hoped to have Mr. Adam Thompson look over the papers, however he has been on sick report for sometime, and I think it would really facilitate matters very greatly if you will send to me a copy of the agreement as we finally concurred in at the meeting at Major Peterson's home. It may not be necessary for us to see any attorney at all if this paper is sent directly to me first.…I am very anxious, in view of my friendship and appreciation of work which has been done throughout the years by our faculty members who have become interested, to get the corporation completed as soon as possible."

The Adjutant, 1937, yearbook published by the cadets of the San Diego Army and Navy Academy.

Chapter 9: The Interregnum

From "History on Select Presidents of the Army & Navy Academy," circa 2003: "[John Lynch Davis] proposed the construction of a new central building.…The idea appealed to the Andersons, and in 1938, the building was erected and paid for 'cash on the line' by [them].…Major Davis [later] accepted an attractive offer to move to an eastern school."…"Captain V.R. Vestal, a graduate of West Point. Because his term of office extended little more than one year, he is usually listed as an 'acting president.' He has the distinction of being the first President of the Academy when it became a non-profit charitable and educational organization.…The non-profit incorporation of the Army and Navy Academy, Inc. was accomplished on December 16, 1941."…"Mr. and Mrs. Anderson [had] suggested that the stock company be dissolved—the stock was still valueless—so that the Academy could become a non-profit school, which included special federal tax-exemption benefits…so they voted a cancellation of the stock company and supported the establishment of the Academy as a non-profit charitable educational organization."

Chapter 10: To Same Heights

From "The Army & Navy Academy: Its Story 1910 through 1936," by Colonel William C. Atkinson, 1972: "They [the Atkinsons] decided that they would prefer a larger military school, preferably in Southern California, this was in the spring of 1924.…They arrived in San Diego on September 24th and after much inquiry found their way to the school only to be met with great disappointment.…As they climbed the front steps they met Major John Davis, the Colonel's younger brother.…[T]hey had found him so friendly.…[T]hey accepted the challenge to apply themselves with everything they could muster."

From Robert Walker Tribute website and Baskette, Kirtley. "Bob Walker's Life Story." *Modern Screen* (January 1946): "Walker grew up on Ogden, Utah. At San Diego Army and Navy Academy learned much needed discipline. Head of Dramatics Department, Virginia Atkinson, was impressed with his musical abilities on drums and transferred him to drama class."…"That's how Bob Walker found himself, next school season, enrolled as a 'rat' in the San Diego Army and Navy Academy, in Southern California. At the start, he hated the place. At San Diego you ate, dressed, studied, played and slept to bells and bugles. He was to live in a barracks with another roommate, wear a uniform modeled after a West Point cadet's, was to carry himself like a ramrod and drill like a wooden soldier."…"Atkinson changed his life forever.…He wrote to family that night: 'Dear Mother, I guess tonight I am the best young actor in all San Diego.' and he went on from there. At the bottom he penned, 'send this on to Aunt Tenny.' He got letters back from them all. They were proud. They knew he had the stuff, and right away Bob began proving it."

Quotes from Robert Walker from Robert Walker Tribute website and Baskette, "Bob Walker's Life Story": "[I]t's life for you that I should wind up at the San Diego Army and Navy Academy, where discipline IS discipline!"…"I had been at the military academy about a year, and hating every part of it, when one day Mrs. Atkinson asked me to read. She was a kind and wonderful woman, and she must have been a really fine psychologist, too, because she got me so interested in dramatics I forgot how much I hated discipline. For the first time in my life, my future came into focus. From that day to this one, I have had just one ambition—to be a good actor. What would have happened if she hadn't seen something in me other people did not, what would have happened if I hadn't taken that first course in drama, I don't know. I think that perhaps my whole life might have been changed. Of course, it helped that I thought she was the loveliest woman in the whole world!"

From Canfield, Alyce. "You Can't Do That, Bob Walker." *Screenland* (January 1945): "When anyone asks what my first 'break' was, I always say, 'Meeting Mrs. Atkinson.' I've always kept in touch with her, and I hope someday she'll be proud of me."

Part IV: The Atkinson Era

Chapter 11: Atkinson's Lament

Most of the information in this chapter comes from Howard-Jones, Marje. *Seekers of the Spring: A History of Carlsbad.* N.p.: Friends of the Carlsbad Library, 1982, and the *Carlsbad Journal.*

From "Let's Get Acquainted," an advert created by Colonel Atkinson promoting the Academy, speaking in first person from the Academy's perspective: "I should tell you that I did much to place Carlsbad on the international map through my magazine advertising, and, as a result, I draw boys from many foreign countries. Even today, I suppose I do more to bring Carlsbad to the attention of the international public than any other agency in our area. I should also tell you that one Sunday each month I hold interviews for parents…and this brings to Carlsbad patrons.…I hold weekend events that bring patrons to Carlsbad for the weekend, and this brings business to motels, restaurants, service stations and other businesses in general."

Chapter 12: Towards the Sunset

Most of the information in this chapter comes from Bitterlin and Radovich, *Four Academies*; catalogues from Brown Military (circa 1940s to 1950s); John Brown University website; catalogues from Brown Military Academy; and the Brown Military Academy Alumni Association website.

Other information came from interviews made with alumni of Brown Military Academy–Pacific Beach, primarily Doug Halbert (Class of 1950, adjutant of the BMA Alumni Association).

From Bitterlin and Radovich, *Four Academies*: "When Brown closed in 1958 the last Battalion Commander to lead the Corps was Melvin Stein, now a successful San Diego attorney. Also present at the last Dress Parade ever to be held at Brown Military Academy in June 1950 was Colonel Davis who was invited by Bitterlin, then Superintendent, to be the Reviewing Officer. Although too fragile to 'troop the line,' the Colonel stood erectly to return the salute of the Battalion Commander. It was a poignant scene: the Colonel watching as the Cadet Corps marched off the field for the last time."

From The San Diego Community News Group—Now & Then website: "50 Years Ago: The 'Last Parade' at Brown Military Academy took place on Wednesday, June 10, 1958. Honorary Reviewing Officer was Col. Thomas A. Davis, who founded the school as the San Diego Army & Navy Academy in 1910. 'Davis tried valiantly to stand to take the review,' said the *Evening Tribune*. '(Gen. John) Harmony gripped his right elbow to steady him, but he couldn't make it.' The school population would move the next year to Monrovia, and the buildings would be torn down to make way for Pacific Plaza Shopping Center."

Interview of Brown Military Academy alumnus Leonard Cordoza (Class of 1941), conducted by Michael D. Milligan and Sharon Milligan, May 12, 2011, available at the Brown Military Academy Alumni page.

Chapter 13: The Golden Anniversary

From the fiftieth anniversary edition of *The Warrior* published in 1960. "1960—Senior White Uniform debuted as part of Golden Anniversary Celebration."…"1961—Academy officials surprised when it was announced San Diego Chargers considered ANA as a site for possible summer training site, they weren't contacted."

Fiftieth anniversary letter by Colonel Atkinson in *The Warrior*: "Fifty years is a long span of years in the life of an individual or an institution. Both are alike, however, as related to the ultimate end of the span in so far as qualities of success, dignity, and ideals are concerned. Of the fifty years of the Academy's life it has been my privilege to have served it in one capacity or another for all but fourteen of the fifty years."

Blade-Tribune. "Carlsbad Army-Navy Academy Looks Back on Good, Bad Times on 50th Anniversary" (November 7, 1960).

San Diego Union. "Army-Navy Academy to Observe 50th Anniversary." November 7, 1960.

"History on Select Presidents of the Army & Navy Academy," circa 2003: "Every piece of chinaware had been fired with a red apple design. With no money to replace the china, however, the group felt it wise to allow the design to vanish through attrition. Many years later, the group celebrated as the last piece of 'red apple' china plate was ceremoniously broken."

Craven, Stephanie. "Red Apple Inn Nothing but a Tasty Memory." *San Diego Union-Tribune*, September 22, 2002.

Information on the red apple story comes from an interview with Joe Orozco, conducted in 2006.

Information on the mascots comes from Burden, John. *Army Navy* 5, no. 2. "ANA Canine Mascots: Cadet's Best Friend" (Spring 2010).

Donald H. Cline, 1957, West Point Memorials. http://apps.westpointaog.org/Memorials/Article/21393.

PART V: THEY JUST FADE AWAY

Chapter 14: All Glory Is Fleeting

The chapter title "All Glory Is Fleeting" is a reference to the final line of the *Patton* movie, which was required viewing in LET during my time at the Academy.

Information on Brown Military Academy–Glendora came from e-mail and phone call interviews made with alumni, primarily Gary Moore (Class of 1963).

Interview of BMA alumnus Jim Griffith (Class of 1963), available at the BMA Alumni page.

Information on San Diego Military Academy primarily comes from Bitterlin and Radovich, *Four Academies*, and the San Diego Military Academy Alumni Association website and Facebook group.

Information on Elsinore Naval and Military School comes from the ENMS Alumni website and Rokos, Brian. "Lake Elsinore: Owner Vows to Maintain Military Academy." *Press Enterprise*, March 1, 2012.

Williams, Jack. "Louis Bitterlin, 88, Pacifist Founded San Diego Military Academy." *San Diego Union-Tribune*, February 26, 2003.

Chapter 15: Strange Sense of Nostalgia

Most of the information in this chapter comes from the document "History on Select Presidents of the Army & Navy Academy," circa 2003: "It didn't take long for Colonel Cline to realize why the campus appeared to be the same as it had been for the past fifteen years, and indeed, for many prior to 1968. The small Board of Directors at that time consisted of three civilian individuals.... These three held tight reins on the operation of the Academy and completely controlled the purse strings. Colonel Cline became the fourth member of the Board, but the voting was almost 3-1 against the President on any issues requiring expenditure of funds for the betterment of the operation of the Academy."..."Soon after arriving at the Academy in fall of 1983, Colonel Cline began to attend meetings of the [AMSCUS]....These organizations provided him a wealth of information about how peer schools were being run and what improvements could be instituted....[A]rmed with many new ideas; Colonel Cline carefully drew up plans and began to present ideas to the Board....Most of his suggestions for change fell on deaf ears."..."A major problem at that time was that the starting salary for a faculty member at ANA was less than half of the starting salary of a public school teacher in [the] area. Another example: the Cadets were taking typing classes using antique manual typewriters....During the time Colonel Cline served as President, salaries of the faculty increased by approximately 80 percent, and salaries of support staff increased approximately 50 percent...but this was accomplished only after painful annual battles with the Board."..."These improvements, just to mention a few, came at the expense of many uncomfortable meetings with the very conservative Board of Directors. Board meetings usually took place in Thomas Sefton's office in San Diego with the three civilian Board members seated comfortably on an expensive couch, with Colonel Cline seated directly before them in an uncomfortable wooden chair. At one point the Board ordered Colonel Cline not to wear his military uniform to Board meetings anymore because it made them feel too intimidated!"..."On the morning of January 16, 1991, Colonel Cline received an abrupt telephone call from Vroman Dorman [who informed him that]…'a decision had been made to terminate you, Colonel Cline, as President; that Dr. Conrad Garner and his wife are coming in immediately to take over; and a special meeting of the Board of Directors is scheduled for January 25, 1991, at 9 a.m. in the offices of Thomas Sefton, at which time formal action to terminate Colonel Cline will be made.'….All of this came as a complete and sudden surprise to Colonel Cline."

Yasuda, Gene. "Academy Rewards: Even Cadets Not Here by Choice Salute This School." *Los Angeles Times*, April 30, 1989.

Information on the Southern California Military Academy came from e-mail and phone call interviews made with alumni of Southern California Military Academy, primarily Roger Van Rekom (Class of 1948), and the SCMA Facebook page.

Appleman, Marc. "Chasing a Legend Named Dad: Army and Navy's Jason Simpson Is Making O.J. Proud." *Los Angeles Times*, October 10, 1986.

Rogers, Thomas. "Scouting, All Over Again." *New York Times*. October 10, 1985.

Theories on Jason Simpson and the *People v. O.J. Simpson* case come from Dear, William. *O.J. Is Innocent and I Can Prove It.* N.p.: Skyhorse Publishing, 2012.

Chapter 16: The Year of Five Presidents

Author's note: I apologize if this chapter looks sparse and disjointed, as large cuts and edits have been made from the earlier drafts of the book shortly before the manuscript was submitted for publication. This was a mutual decision by numerous parties, as the original version of this chapter focused more on sole individuals, which took away from the overall narrative of focusing on the Army and Navy Academy as a whole.

Most of the information in this chapter came from interviews with alumni and staff members of this era, the 1992 *Adjutant* yearbook and the document "History on Select Presidents of the Army & Navy Academy," circa 2003: "Conrad Garner and his wife arrived on campus….During a stormy month or two, he met with the faculty and ordered everyone to submit their resignations. Mrs. Garner stayed on campus while Dr. Garner returned to their previous residence to retrieve household goods…within a matter of hours both Garners disappeared as suddenly as they had arrived. It was never announced whether they left on their own accord, or whether they had also been terminated by the Board."

Other information came from an interview with Johnnie Crean, conducted in November 2014. Johnnie Crean served on the Board of Directors of the Army and Navy Academy from 1991 to 2003. Previously, Johnnie was known for working with his parents' RV company, Fleetwood Enterprises, and for setting a record for the amount of money spent on campaigning for a House seat, $2.7 million, as reported by the *New York Times* in 1982. In that election, he went on to lose to a write-in candidate, Carlsbad mayor Ronald Packard, also setting the record for biggest marginal loss to a write-in candidate—Packard saw the accomplishment of being the first write-in candidate to defeat the candidates of the two major parties. Parts of the interview I conducted with Johnnie Crean are presented here: "[The Board was no] longer speaking with each other. So anyways they had a meaning called me, in San Diego, elected me as the 5th person, I recommended they form an operations committee, so they had a one person operations committee, that was me, then they adjourned the meeting then left. So from that point on for about a decade I ran the school."…"I made a list of the staff so I went through and I talked to all the members of the staff and the BC and some of the cadets, and I had the names of all the employees on one side of the paper and on top that I talked to on the list, it was like a grid. I let everyone discuss everybody, so I end up reducing, letting a third of the people go, I didn't know anyone, but I fired based on the marks."…"Hired Miller since he was brought up in ROTC and had respect for the academic side, Board shacked Heinle then formed the Office of the President, [Crean] was the CEO in charge of it, inside the office was [three] the Superintendent, Dean of Academics, and CEO, all ran under [Crean's] supervision, school ran like that from when Heinle left until a year before Bliss came."

Turner, Wallace. "Californian's Plan to Outspend His Foes Backfires." *New York Times*, September 8, 1982.

Pasco, Jean. "Suit Filed in Leadership Dispute at Military School in Carlsbad." *Los Angeles Times*, April 13, 2003.

Mickadeit, Frank. "Johnnie Crean: Parrot-Head or Air-Head?" *Orange County Register*, August 21, 2013.

Part VI: The Academy's Renaissance

Chapter 17: Into a New Millennium

Most of the information in this chapter comes from interviews and e-mails with Karen Chemlir Miller, wife of the late Lieutenant Colonel Stephen

Miller: "Early in his position as Senior Army Instructor (1991), Col. Miller realized the need for revitalizing a culture of tradition when he discovered that few seniors knew the words to the Alma Mater. From that day forward, singing the Alma Mater became a prerequisite for passing leadership training courses, and the motto 'Warrior Pride' became woven into all aspects of ANA campus life that continues to this day."

Chapter 18: Foundations of the Revival

Most of the information in this chapter comes from statements and letters by General Stephen Bliss in the 2004–8 issues of *Army Navy* magazine, as well as an interview conducted with Bliss in 2006.

Information on Brigadier General Stephen M. Bliss's 2004 term comes from *Army Navy* 1, no. 1 (Fall 2005). Quotes from Bliss come from *Army Navy* 7, no. 2. "History in the Making" and "Hail & Farewell, BG Stephen M. Bliss" (Winter 2014).

Chapter 19: Brother Stands by Brother

This chapter purposely makes no use of any public sources or publications as a stylistic choice, as it is more of an autobiography of the author and is meant to capture recollections and memories of the cadet experience firsthand at ANA.

Quotes from First Sergeant Martin Davis come from the Facebook page "Watching 1st Sergeant Davis Yell at People During Inspections."

Part VII: Towards the Centennial

Chapter 20: An Industrial Age

Army Navy 5, no. 1: "Raise the Flag for Army Navy" and "USS Stockdale Officers and Crew Cheer Warriors to Victory" (Fall 2009).

Chapter 21: Centennial Celebrations

Much of the information in this chapter comes from public press releases, articles and advertisements done by ANA to promote the Centennial Celebration. These sources are easily accessible to the public.

Holeman, Angela. "Academy Celebrates Centennial Year." *Carlsbad Business Journal*, January 2010.

North County Times. "Centennial Celebration Weekend." October 6, 2010.

Information on the Veteran's Memorial and Centennial Weekend comes from *Army Navy* 6, no. 1. "A Fitting Tribute to Academy Veterans" and "Party of the Century—Centennial Gala 2010" (Fall 2010).

Information on the Warrior Band in the National Memorial Day Parade comes from *Army Navy* 7, no. 1. "National Memorial Day Parade: A Special Addition to the Academy Centennial!" (Fall 2011).

Chapter 22: Into a New Century

Much of the information in this chapter comes from public press releases, articles and advertisements done by ANA. These sources are easily accessible to the public.

Duffield Sports Center opening comes from *Army Navy* 7, no. 2. "Duffield Sports Center Opens to Enthusiastic Cheers" (Winter 2014).

Information quoted in this chapter comes from *Army Navy* 7, no. 2. "First 100 Days of Presidency," "The Gurian Institute Is Coming to the Academy!" and "Trip of a Lifetime—Cadets Travel to Normandy" (2014).

Military School Enrollment Reports denote the following as of 2014:

- Oak Ridge Military Academy, enrollment of 45 in 2014. Tiberii, Jeff. "16-Year-Old Oak Ridge Military Academy Marches On." *WUNC 91.5 Public Radio*, September 4, 2014.
- Howe Military Academy, enrollment of 74 in 2014. Peterson, Mark. "Possibility Is 'Very Strong' that Howe Military Academy Will Close." *16 WNDU-TV*, May 19, 2014.
- Chamberlain-Hunt Academy, enrollment of 4 in 2014. *WAPT 16 News*. "Chamberlain-Hunt Academy to Close." July 30, 2014.

Select Bibliography

The Adjutant. Yearbook of the Army and Navy Academy. Carlsbad, CA. Years 1938, 1939, 1956, 1960, 1965, 1971, 1991 and 2008.

Army Navy. Magazine of the Army and Navy Academy. Carlsbad, CA. Spring 2008, Fall 2009, Spring 2010, Fall 2010, Spring 2011 and Winter 2011.

Atkinson, William C. "The Army & Navy Academy: Its Story, 1910 through 1936." Carlsbad, CA, 1972.

Bitterlin, Louis T., and Milan Radovich. *The Four Academies.* Carlsbad, CA: self-published, 1996.

Blade-Tribune. "Carlsbad Army-Navy Academy Looks Back on Good, Bad Times on 50th Anniversary." November 1, 1960.

Brookhart, Robert. "A&N Has Colorful Past." *The Warrior*, May 1961.

———. "Building Program Covers 25 Years." *The Warrior*, September 1960.

———. "Presidents and Headmasters of the Academy." *The Warrior*, November 1960.

The Cadet. Student newspaper of the San Diego Army and Navy Academy. Articles published in 1912–13.

The Cadet. Yearbook of the Davis Military Academy. Carlsbad, CA. Year 1937.

The Cadet. Yearbook of the San Diego Army and Navy Academy. Pacific Beach, CA. Years 1912, 1913, 1914, 1917, 1929, 1930, 1931, 1935 and 1936.

Canfield, Alyce. "You Can't Do That, Bob Walker." *Screenland* (January 1945).

City of San Diego and San Diego County. Vol. 1, *The Birthplace of California*. San Diego, CA: American Historical Society, 1922.

"History on Select Presidents of the Army & Navy Academy." Carlsbad, CA, 2003.

Howard-Jones, Marje. *Seekers of the Spring: A History of Carlsbad.* Carlsbad, CA: Friends of the Carlsbad Library, 1982.

Interviews with Steven M. Bliss, 2006; John Maffucci, 2006; Joe Orozco, 2006; Johnnie Crean, November 2014; and Gary Moore, November 2014.

McGroarty, John S. *Los Angeles from the Mountains to the Sea.* Vol. 3, *Select Biography of Actors and Witnesses to the Period of Growth and Achievement*. Los Angeles, CA: American Historical Society, 1921.

Ramirez, Samuel A. *The Junior Historian: El Paso Military Institute.* Austin: Texas State Historical Association, 1954.

Ryden, Roy. "Celebrate 50th Anniversary of ANA." *The Warrior*, May 1961.

San Diego County California. Vol. 2, *A Record of Settlement, Organization, Progress and Achievement.* Carlsbad, CA: S.J. Clarke, 1913.

Sargent, Porter. *A Handbook of American Private Schools*. 7th ed. Sargent's Handbook Series. Boston: Cosmos Press, 1922.

The Warrior. Student newspaper of the Army and Navy Academy. Articles published 1960–61.

Williams, Jack. "Louis Bitterlin, 88, Pacifist Founded San Diego Military Academy." *San Diego Union-Tribune*, February 2003.

Yasuda, Gene. "Academy Rewards: Even Cadets Not Here by Choice Salute This School." *Los Angeles Times*, April 30, 1989.

Index

A

Adjutant, The 106
All Southwestern Players 91
alma maters (anthems) 83, 112, 121, 142, 198, 213, 226
alumni 14, 48, 56, 75, 95, 125, 141, 145, 148, 150, 166, 172, 176, 178, 198, 202, 208, 212, 216, 223, 229, 232, 233, 236, 237, 240, 243, 247, 251
Army Navy magazine 201, 231
Association of Military Colleges and Schools of the United States (AMCSUS) 195, 199, 217, 246, 247
Atkinson, Virginia 73, 88, 101, 110, 112, 115, 125, 137, 145, 148, 153, 198, 220
Atkinson, William C. 23, 72, 83, 92, 101, 105, 107, 109, 110, 116, 119, 125, 128, 131, 143, 148, 150, 153, 154, 160, 163, 167, 169, 174, 178, 185, 198, 220, 237, 240

B

Bartell, Arthur M. 246, 249
basketball 46, 50, 91, 143, 162, 165, 167, 210, 224, 243, 245
Beadle, Jesse A. 41
Bitterlin, Louis 23, 139, 144, 150, 161, 164, 172
Black-Foxe Military Institute (Hollywood, CA) 74, 75, 167
Bliss, Stephan M. 201, 218, 238, 242, 244, 247
Board of Directors 107, 115, 148, 165, 171, 180
Board of Trustees 89, 93, 175, 204, 244
Brookhart, Gilbert N. 167, 170
Brown, John 101, 104, 129, 134, 137, 140, 144, 160, 162, 173
Brown Military Academy 24, 104, 119, 129, 134, 137, 141, 143, 147, 150, 162, 164, 168, 172
Brown Military Academy of the Ozarks 142, 144, 160
Brown School for Girls 137, 160, 172
Burbank Military Academy 74, 75

C

Cadet, The 38, 45, 70, 91, 94, 106
California Military Academy 75, 161
Camp Pacific (Academy by the Sea) 118, 183

Carlin Military Academy 171
Carlsbad 23, 95, 99, 102, 104, 106, 108, 109, 113, 115, 118, 119, 126, 128, 131, 134, 139, 143, 145, 148, 150, 152, 154, 162, 166, 171, 183, 202, 207, 211, 222, 237, 244, 250
Castle Heights Military Academy (Lebanon, TN) 27, 110, 168
Charlotte Hall Military Academy (Charlotte Hall, MD) 17, 168
Cline, Donald H. 154, 169, 174, 178, 180, 183, 201, 205, 237

D

Davis, Bessie Taylor 26, 57, 58, 205
Davis, John Lynch, Jr. 25, 58, 105
Davis, John Lynch, Sr. 25
Davis Military Academy 95, 100
Davis, Thomas Alderson 23, 25, 27, 31, 35, 38, 40, 46, 48, 49, 52, 55, 59, 61, 64, 69, 72, 74, 78, 82, 87, 89, 93, 94, 100, 101, 104, 105, 108, 110, 125, 129, 134, 138, 140, 143, 145, 150, 152, 163, 170, 185, 196, 238, 250
Dorman, Vroman James 102, 104, 116, 125, 146, 159, 171, 180, 184
drill teams 166, 171, 196, 198, 209, 210

E

Ede, Raymond C. 119, 129, 131, 132, 133, 203, 237
El Paso 27, 31, 36, 48, 58, 59, 152
El Paso Military Institute 28, 31, 36, 40, 47, 48
Elsinore Naval and Military Academy 74, 168
Emery, Grenville C. 74, 75, 78

F

football 46, 48, 49, 72, 83, 85, 90, 106, 121, 142, 162, 165, 167, 176, 179, 183, 208, 218, 230, 236

G

Garner, Conrad R. 180, 183
Girl's Collegiate Institute 137, 160
Greenbrier Military Academy (Lewisburg, WV) 110, 168

H

Harding Military Academy (Glendora, CA) 171
Harvard School for Boys (Los Angeles, CA) 75, 168
Heinle, Fred W. 185, 186, 197, 199, 203
Henry, Mrs. John H. 60, 61, 64
Hollywood Military Academy (Hollywood, CA) 74, 75, 77
Hotel Balboa 35, 38
Huntington Hall (Pasadena, CA) 76

J

John Brown Schools 135, 141, 160, 173
John Brown University (Sulphur Springs, AZ) 139, 141, 144, 160
Junior Reverse Officers' Training Corps (JROTC) 14, 16, 17, 74, 78, 118, 154, 171, 172, 194, 195, 203, 205, 215, 218, 222, 233, 246, 247

L

Leasure, Derrick 203, 231, 233
Long Beach Military Academy (Long Beach, CA) 75, 77

M

Maffucci, John 143, 179, 183, 208
Manzanita Hall (Palo Alto, CA) 74, 168
marching bands 31, 72, 112, 213, 239
Miller, Stephen A. 193, 197, 200, 201, 203, 205, 237, 238

N

Norwich University (Norwich, VT) 19, 20

O

Oneonta Military Academy (Pasadena, CA) 76

P

Pacific Beach 35, 38, 49, 54, 58, 60, 62, 73, 76, 83, 90, 93, 101, 102, 105, 108, 110, 113, 119, 129, 131, 137, 139, 140, 143, 145, 147, 150, 152, 160, 162, 163, 172, 220, 235, 237
Pacific Military Academy (Culver City, CA) 74, 77
Page Military Academy (Los Angeles, CA) 77
Palo Alto Military Academy (Palo Alto, CA) 74, 75, 168
Park Military Academy (Palo Alto, CA) 74, 75
Partridge, Alden 18, 19
Pasadena 46, 58, 60, 62, 64, 72, 74, 76, 82, 92, 134, 152
Pasadena Army and Navy Academy (Pasadena, CA) 59, 62, 64, 173
Pasadena Military Academy (Pasadena, CA) 63, 64, 76
Peterson, Samuel Warfield 57, 83, 107, 146, 203

R

Reserve Officers' Training Corps (ROTC) 19, 22, 24, 79, 107, 173, 195, 208, 246, 247
Rose Bowl (Pasadena) 72, 83, 112
Rowbotham, Fred 170, 180

S

San Diego Army and Navy Academy (Pacific Beach, CA) 14, 24, 37, 38, 40, 49, 54, 59, 62, 70, 72, 83, 87, 90, 94, 101, 104, 107, 110, 113, 115, 129, 134, 137, 142, 145, 152, 160, 163, 166, 205
San Diego Army and Navy Junior College (Pacific Beach, CA) 80, 83, 84, 91, 139, 162, 248
San Diego Military Academy (Solana Beach, CA) 23, 163, 166, 168, 172
Simpson, Jason 179
Southern California Military Academy (Long Beach, CA) 74, 75, 77, 142, 162, 168, 172
South Western Military Academy (Long Beach, CA) 74, 77, 172
St. Catherine's Academy 22, 74, 168, 173
Strip, the 129, 131

U

United States Military Academy (West Point, NY) 18, 20, 24, 29, 54, 107, 201
Urban Military Academy (Hollywood, CA) 74, 75

V

Vestal, V.R. 107, 201
Virginia Military Institute (Lexington, VA) 20

W

Walker, Robert Hudson 113, 115
Warrior, The 105, 125, 147, 149, 178
West Coast Military Academy (Palo Alto, CA) 74, 75
Wood, Charles Milton 43, 45, 50, 61, 63

Y

Young Men's Christian Association (YMCA) 26, 30, 37, 40, 50

ABOUT THE AUTHOR

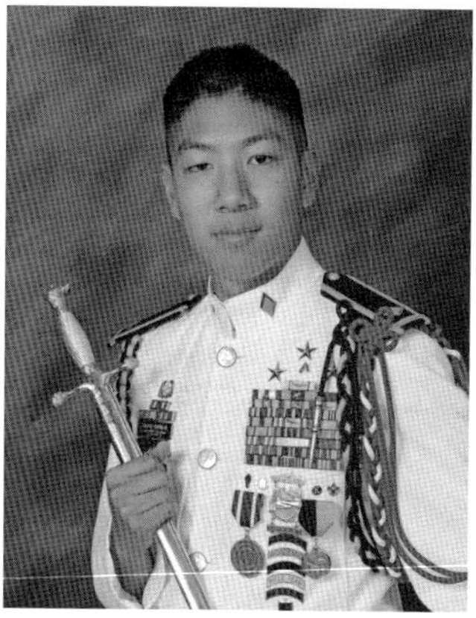

Courtesy of NGM Photography.

Alexander Mui is a top graduate of the Army and Navy Academy, where he served as Battalion Executive Office and Alumni Class agent. He helped reestablish the Academy's museum and documented the first complete history of the school from more than a decade of research. He attended Johns Hopkins University on the Hudson Trust Scholarship, where he studied for a Bachelor of Science degree in molecular and cellular biology, while at Hopkins he served as a senior editor for the award-winning *JHU News-Letter*.